融合与发展
景观统筹一体化建设

风景融入日常生活

谢晓英　张琦　周欣萌　刘晶·著

化学工业出版社

·北京·

图书在版编目（CIP）数据

融合与发展：景观统筹一体化建设/谢晓英等著.
—北京：化学工业出版社，2023.2
（风景融入日常生活）
ISBN 978-7-122-42595-9

Ⅰ.①融… Ⅱ.①谢… Ⅲ.①城市景观-景观设计-北京 Ⅳ.① TU984.21

中国版本图书馆CIP数据核字（2022）第230581号

本书翻译（中译英）

Wang Yile（澳大利亚）　　Daniel Lenk（英国）

责任编辑：林　俐　刘晓婷　　　　　　　装帧设计：对白设计
责任校对：王　静

出版发行：化学工业出版社（北京市东城区青年湖南街13号　邮政编码100011）
印　　装：北京宝隆世纪印刷有限公司
710mm×1000mm　1/12　印张16　字数250千字　2023年3月北京第1版第1次印刷

购书咨询：010-64518888　　　　售后服务：010-64518899
网　　址：http://www.cip.com.cn
凡购买本书，如有缺损质量问题，本社销售中心负责调换。

定　价：98.00元　　　　　　　　　　　　　　　　版权所有　违者必究

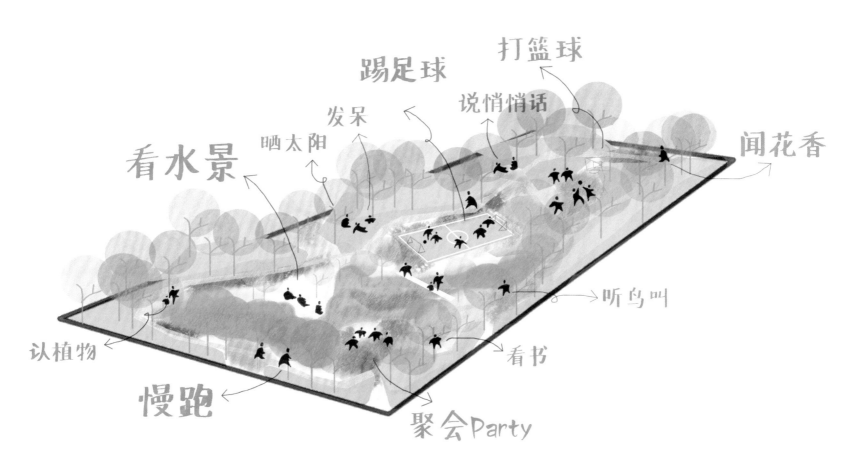

为无界景观设计团队的作品所作序

建筑、景观设计之为专业，不但有其功能意义，更关系一城一地的外部形象。在社会经济快速发展的时期，无疑处于"前沿"。景观设计的成果一旦落地，即成为项目所在地的日常现实，人们生活、呼吸其间，直至习为常态，日用而不自知。无论设计成果存在时间久暂，一种专业活动如此深地进入人们的生活，足以令人称羡。这一专业的"改变中国"，实实在在，不仅诉诸感官（不惟视觉），还影响人们的心态乃至生存状态。纵然有一天设计成果被时间抹去，其印在地面上尤其人们记忆中的痕迹，也是难以消失的吧。

景观设计与建筑设计，均为重塑空间的艺术：由物理空间到人文空间。景观设计师以大地为画布。无界景观设计团队的"绘画"作品散落于祖国大江南北以至不同的国家。团队一以贯之，从唐山凤凰山公园复建，到北京大栅栏杨梅竹斜街的改造，从北京城市副中心行政办公区核心区，到埃塞俄比亚首都亚里斯亚贝巴的友谊广场，均力求因地制宜、就地取材，低能耗，对原有的自然与人文环境少扰动，使人工与自然有机对接，满足多方面的功能需求。

团队在恪守专业人员的工作伦理的同时，不忘尽社会责任。团队的主要贡献在城市建设，却对"乡建"怀有志愿者般的热情；在城市作画之余，更将笔触伸向乡村。近年来"历史文化街区""特色小镇"的建设一哄而起，以改造、升级为名，造成"千村一面"的不可逆的破坏，至于"打造景点"则往往出于政绩冲动，这样的现象亟待规范。优先考虑在地居民的需求，而非将他们的环境连同生活作为展品，强迫其为拉动地方经济做牺牲——景观改造要这样才能永续。

中国的发展为景观设计专业提供了机遇。无界设计团队有幸参与了这一进程，年轻设计师积蓄的能量得以释放。尽管行政力量的干预在所难免，人们环境意识以至社会审美风尚的进步，毕竟是更强大的力量。上述变化，由无界景观设计团队历年的项目即不难窥见。

于"隐身"——即不炫技、不刻意打造设计者自身形象——之外，这个团队贯穿始终的理念，更有"共享"。唐山凤凰山公园设计建造方便市民穿行的步道，赣西南夏木塘项目构建供乡民交流的公共空间，埃塞俄比亚友谊公园满足由官方仪式到民众日常娱乐的诸种需求，使广场成为便于人们交流融合的平台，他们的项目无不以社会效益为重要考量。

无界设计团队首先着眼的是项目所在地民众的日常感受；不取一时亮眼，而求可持续，可再生，可不断更新换代。在我看来，低调的美或更能持久。近年来追求"博眼球"，亮丽过后顿成鸡肋的项目比比皆是。回应民众需求，具备自我更新的能力，维持景观的持久活力，是对治此病的药方。无界设计团队还将改善当地民众的生存状况纳入优先考量，确保公共财政取之于民，用之于民，切实提升在地居民的幸福感。经历近几十年一波波的造景运动，团队的上述追求倍加可贵，值得阐发、揄扬。

长期实践中，团队形成了自己成熟的设计语言，如绿道、步道、步行网络的设计，又如对无机械健身，即"软性的、温和的、能够随时随地进行的健身"的推广。凡此，无不力求最大限度地利用空间，满足多方面的需求。设计团队专业技术与人文层面并重，不但尊重自然，而且顺乎环境的历史脉络。对场地破拆材料进行再利用，既节约能源，又有效地保存场地记忆；以艺术化的地面、立面铺装，借助物料上的时间刻痕保存历史信息，都是团队行之有效的设计手法。砖石往往是既可视又可触的历史。镶嵌在地面、墙体的符号化的历史，以其物质形态嵌入了现实，不同时空的融汇借此种细节显现。用有年代感的砖、不同质料的石材，拼贴成图案承载记忆，这一设计手法在杨梅竹斜街改造、一尺大街修复、北京城市副中心行政办公区的设计与施工中，均有成功的运用。北京城市副中心行政办公区核心区步行道边镌刻北京新老胡同名，小广场铺设嵌入取自胡同的石材、木料，无不将老北京的文化密码不张扬地嵌入墙体、地面，只待行走者辨识。设计团队更注重利用现有的科技手段，实现"历史文化信息库的搭建，实现线上线下空间交叠"，既丰富了人们的空间体验，又使民众于娱乐休闲的同时摄取历史知识。凡此种种，随处可感用心之细，用情之深。

较之纸上的画作，实现在空间的绘制更难抵抗时间的侵蚀。景观设计作品势必经历一轮轮的更新、升级，也由此重生。原有设计中的亮点，有可能烛照、开启以后的设计思路。这也是设计团队为后续的"创作"预留空间的必要性及意义所在。

我一再说自己是书斋动物，因此对所有切切实实促进现状改善的努力都怀有敬意，对于景观设计专业不得不应对的诸种难题略知一二："设计"不能止于图纸、文案，须将设想落实到施工现场，以至监督、干预用料的制作；具体实施中更要与相关各方磨合，做不得已的妥协。所有这些，岂是我这样的书生所能应对！作为行外人士，我欣赏的，毋宁说更是无界设计团队的理念，即如"安住""乐活"，另如"微更新""微改造""微设计""轻介入""轻资产运营""低成本开发""小尺度下的微改造"，以至"如针灸一般微创与介入式的改造"——切中时弊，较之具体项目，或许更有推广的价值。对成效的估量或有不同，上述原则对于相关行业，值得一再重申。无界设计团队所实行的"微改造"，或许多少也因财力所限，但也不妨换一种思路：这也是在严格给定的条件下创造最大的效益。

无界设计团队一再扩大作业范围，由京城而外地而乡村而国外。项目对象有北京城市副中心，也有村落；项目性质有外地商业性质的，也有带有公益性质的京城旧城区改造，以及援外项目，无界设计团队无不从容应付，并因项目性质的变换、场域的转移、涉及范围的扩展而自我提升、完善，化蛹成蝶。尽管岁月轮转，时间推移，这一团队长时间地保持生机勃发，展现出不会枯竭的原创性与潜能。我有幸或多或少地见证了这一过程，领略了团队虽受困于现实条件但仍不放弃专业理想与职业操守的顽强，也由他们坚持的理念、思路获得滋养。

对于无界设计团队业绩的评估，既有空间的也有时间的尺度。无论怎样，在我看来，正是这种一点一滴的改变，一片区一角隅的重新塑造，影响着未来中国的面貌。纵然因形格势禁，一些富于创意的设计未能实现在地面上，也以设计图、文案的形式为行业提供借鉴。纸墨更寿于金石，为一个时代的行业状况留文献，岂不是有不可替代的价值？

业界将包括无界景观设计在内的一些设计团队的项目介绍称为"来自前线的报告"，甚得我心。在一轮轮的城市改造、"新农村"建设潮中，建筑、景观设计的确位于"前线"。关心未来中国样貌的人们，无疑希望继续收到这种"来自前线的报告"。

赵园

前言
PREFACE

新型城镇化强调城镇化建设质量内涵的全面提升，从"增量规划"向"存量与减量规划"转型，追求节约集约、审慎高效地利用城市土地，从而解决复杂的城市问题。

21世纪以来，城市发展由传统的外拓、粗放模式逐渐向内涵、集约模式转变，人们愈加关注城市的品质与宜居性。城市绿色空间作为城市中以自然生态为主体的物质空间，其布局、质量与功能是决定城市品质的重要因素。然而，随着城镇化的进程加快，高密度城市可用的土地资源愈加紧张，城市面临因人口增多而日益严重的环境和生态问题，其表现之一即城市绿色空间不断被挤压、蚕食。

2015年9月，联合国可持续发展峰会正式通过17个可持续发展目标，旨在以综合方式彻底解决社会、经济和环境三个维度的发展问题。其中，对于城市提出了"建设包容、安全、有风险抵御能力和可持续的城市及人类住区"的发展目标。在此背景下，与市政工程统筹建设、融合人文及自然要素的城市绿色空间建设，或可成为在"存量与减量规划"背景下，改善城市环境、实现低碳减排、提高城市宜居品质、促进城市可持续发展的重要手段。在此理念下引导的城市建设，将更加注重资源的合理高效利用，注重人与自然的和谐，由此推动社会的良性发展。

在二十余年的景观规划设计实践中，我们一直在探索景观设计作为一种优质媒介，最大限度地发挥整体统筹功能的途径。我们致力于打通专业界限，追求一种

Modern urbanization emphasizes comprehensive improvements in the quality of urban construction and a transformation from incremental planning to stock and reduction planning, by pursuing the economical, intensive, prudent and efficient use of urban land, so as to solve complex urban challenges.

Since the 21st century, urban development has gradually shifted away from a traditionally crude and outward-looking model to a more concentrated and reflective model, which reflects the growing concerns people have about the quality and livability of cities. As a physical space and natural ecology, the layout, quality and functions of urban green spaces are an important factor in determining the quality of a city. However, with the acceleration of urbanization, the available land resources in high-density cities are becoming more and more limited, and cities are facing increasingly serious environmental and ecological problems due to growing populations, of which one of the manifestations is the continuous squeezing and encroachment of urban green space.

In September 2015, the United Nations Sustainable Development Summit officially adopted 17 Sustainable Development Goals (SDGs), which aim to thoroughly address the three dimensions of development-social, economic and environmental-in an integrated manner. One of the development goals for cities is "to build inclusive, safe, risk-resilient and sustainable cities and human settlements". In this context, the construction of urban green spaces, which integrate humanistic and natural elements with municipal projects, play an important role in improving urban environments and urban livability, achieving reductions in carbon emissions, and promoting sustainable urban development in the context of stock and reduction planning. Guided by this concept, urban construction pays greater attention to the rational and efficient use of resources and ensuring harmony between human beings and nature, thus promoting the virtuous development of society.

With more than twenty years of landscape design and planning experience, we are constantly exploring ways to maximize integrated functions into landscape design as a medium for the advancement of society. We are committed to bridging professional boundaries and pursuing a "borderless" paradigm of cooperation, connecting urban features and the natural environment with people's daily lifestyles, needs, perceptions and behaviors,

"无界"的合作范式，将城市设施与自然环境，与人的日常生活方式、需求、观念、行为连接起来，促进城市有机体复合功能的融合发展，使景观设计向更广阔的领域敞开。

景观规划设计统筹是因地、因时、因人而异的解决方案，是出于城市整体空间和功能结构的考虑，以厘清场地基因为前提，从场地使用者的实际需求出发，将城市绿色空间、城市基础设施、文化服务设施一体化设计，创新建设城市绿色基础设施的新形式，使城市交通、城市能源、生态修复等市政工程与园林绿化相结合。既适应城市功能的需要，又避免了过度建设对城市生态环境的进一步破坏。由此探索景观规划设计整合功能的最大可能性，探索社会、经济、生态、文化与美学等综合效益最大化。

在此过程中，我们始终秉持"风景融入日常生活"的设计理念，秉持对情境中人的关照的设计初心，珍视地域人群共有的文化传承，重视景观设计的社会学功能。我们所追求的最终目标是运用专业手段协调人与环境的关系，缓解人的生存压力，守护"活力基因"，建立良性可持续的生活方式，激发民众活力和场地生产力，提升场所中人的幸福感与归属感。不着痕迹地浸润，自然而然地介入，经由改善城市生态改善人们的生存状态，重塑其对环境的感知。这种似无为的有为，正是我们不懈追求的境界。

本书着重记录了无界景观设计团队在北京城市副中心行政办公区景观规划以及先行启动区的景观设计实践，并

promoting the integration and development of composite functions of urban organisms, and expanding the domain of landscape design.

Landscape planning and integrated design is a solution that takes into account the surrounding environment, timeframe, and people. It is a new model for the innovative construction of urban green infrastructure, which considers the overall spatial and functional structure of a city, the integration of urban green spaces, urban infrastructure, and cultural services, and their relation with the actual needs of users, under the premise of clarifying the site's innate characteristics in accordance with local conditions, and the integration of transportation, energy, ecological restoration and other municipal initiatives within the scope of landscape design. It both adapts to the needs of urban functions and avoids further damage to the existing urban ecology from excessive construction. Based on this premise, it explores the maximum potential for integrating functions related to landscape planning and design, and the integration of social, economic, ecological, cultural and aesthetic benefits.

Throughout this process, we consistently uphold the design concept of "integrating landscapes into daily life", with the core value of caring for people in these contexts, cherishing the value of cultural heritage, and attaching importance to the sociological function of landscape design. Our ultimate goal is to use professional methods to reconcile the relationship between people and their environment, alleviate the pressures of human existence, and ensure an overall dynamism, so as to establish healthy and sustainable lifestyles, stimulate people's vitality and productivity, and enhance a general sense of happiness and belonging in those who wish to "abide". It is a natural response to endeavor to reshape people's living conditions and the perception of their environment by revitalizing the surrounding urban ecology. This approach of "affecting change unseen" is precisely the goal we relentlessly pursue.

This book focuses on the landscape design practices of the View Unlimited design team and their application in the planning of the Beijing Municipal Administrative Center Pilot Start-up Area. It retraces the design team's process, starting more than ten years ago with the landscape design of the Bolong Lake Economic Zone Complex (a new urban area construction project) and the landscape planning and design of Jiefang South Road

回溯了十余年前设计团队在天津渤龙湖经济区综合体景观设计（城市新区建设项目）及天津解放南路地区景观规划设计（城市片区有机更新项目）中的思考。天津的两处项目因多种原因未能全部实施，但其中的设计理念内涵一直承袭延续，最终在北京城市副中心行政办公区先行启动区项目中得以实现。

经由总结再出发，自遗憾中收获，在记录中成长，是我们著作本书的目的，也是我们对自己的期许。

中国城市建设研究院无界景观工作室
2022年10月

in Tianjin (an organic urban regeneration project). For various reasons, these two projects could not be fully implemented but their design concepts have been adopted and continued, ultimately resulting in the realization of the Beijing Municipal Administrative Center Pilot Start-up Area.

The purpose of compiling this book is to offer a summary before embarking on a new journey. It is a collection of the harvests from regrets, the records of our growth, and the expectations we have for ourselves moving forward.

View Unlimited Landscape Architecture Studio, China Academy of Urban Planning and Design
October 2022

目录 CONTENTS

北京城市副中心行政办公区先行启动区景观设计 ············ **001**

01 北京城市副中心行政办公区景观专项规划 ················ 004
02 北京城市副中心行政办公区先行启动区环境景观设计 ················ 008
03 北京城市副中心行政办公区先行启动区林荫路网专项设计 ················ 048
04 北京城市副中心行政办公区镜河河道景观专项设计 ················ 062

天津渤龙湖经济区综合体景观设计 ············ **141**

天津渤龙湖经济区综合体景观设计 ················ 142

天津解放南路地区景观规划及公园设计 ············ **153**

01 天津解放南路地区景观专项规划 ················ 155
02 天津解放南路地区中央绿轴（都市绿洲）城市公园设计 ················ 156
03 天津解放南路地区起步区公园景观设计 ················ 169

后记 ············ **178**

北京城市副中心
行政办公区先行
启动区景观设计

LANDSCAPE DESIGN
PRACTICES FOR THE
BEIJING MUNICIPAL
ADMINISTRATIVE CENTER
PILOT START-UP AREA

项目背景

通州是京杭大运河的起点,数百年以来,由通惠河与北京老城连通,一脉相承的不止于城市气质,更有人文风貌。作为畿辅门户、水陆要冲,通州曾是扼守皇城经济命脉的漕运重镇,成就了往来通济的历史,孕育了开放包容的精神。

岁月流转,通州在新的时代迎来角色的转变。2012年,北京市委、市政府明确提出"聚焦通州战略,打造功能完备的城市副中心",明确了通州作为北京城市副中心的定位。这是北京市围绕中国特色世界城市目标,推动首都科学发展的一项重大战略决策。通州将打造集行政、文化、经济、教育等诸多功能于一体的综合性城市副中心,这不仅是调整北京空间格局、治理大城市病、拓展发展新空间的需要,也是推动京津冀协同发展、优化人口经济密集地区开发模式的需要,其建设关系着首都北京的未来发展。

北京城市副中心选址于原通州新城规划建设区,总占地面积约155平方公里,外围控制区即通州全区约906平方公里。副中心行政办公区位于通州新城东部的06片区,东六环路以东、运潮减河与北运河之间,地处长安街的东延长线上。

2018年,北京城市副中心控制性详细规划获批复。副中心的规划与设计以中共中央政治局的指示为指导思想,遵循城市发展规律,牢固树立并贯彻落实创新、协调、绿色、开放、共享的发展理念,坚持世界眼光、国际标准、中国特色、高点定位,以创造历史、追求艺术的精神,构建蓝绿交织、清新明亮、水城共融、多组团集约紧凑发展的生态城市布局,着力打造国际一流的和谐宜居之都示范区、新型城镇化示范区、京津冀区域协同发展示范区,打造"无愧于时代的千年之城"。

2019年1月,北京市级行政中心正式迁入北京城市副中心,北京市委、市政府等部门在副中心行政办公区正式揭牌。

北京市老城区水系与副中心水系一脉相承

北京城市副中心先行启动区与镜河二期工程总图

Background

The Beijing Municipal Administrative Center is located in the former Tongzhou New City planning and construction zone, which covers an area of approximately 155 square kilometers. The Beijing Municipal Administrative Center is located in the east extension of Chang'an Road, in Zone 06 of the eastern section of Tongzhou New City, east of the East Sixth Ring Road, between the Yunchaojian River and Bei Waterway.

The construction of the Tongzhou sub-center is a major strategic decision made by Beijing to promote the scientific development of the capital, with the goal of creating a world-class city with Chinese characteristics. To achieve this, it will be vital to adjust the geographical layout of Beijing, manage "big city disease", and promote the expansion of new spaces for development, which will require the coordinated development of Beijing, Tianjin and Hebei, so as to explore an optimal development model for densely populated economic areas. Beijing's urban sub-center is connected to the old city via the Tonghui River, and boasts an urban atmosphere that also draws from this rich cultural lineage. As a comprehensive urban sub-center that integrates administrative, cultural, economic and educational functions, its construction is closely tied to the future development of Beijing and the Beijing-Tianjin-Hebei region.

01 北京城市副中心行政办公区景观专项规划
Beijing Municipal Administrative Center Landscape Design and Planning

- **项目地点：** 中国 北京市通州区
- **项目规模：** 约 6 平方公里
- **设计时间：** 2016—2018 年

- **Project location:** Beijing, China
- **Project scale:** 6 square kilometer
- **Design period:** 2016-2018

从农业时代到工业时代，再到知识经济时代，知识的获取、信息的交流促使人们的生活不断变革。作为"信息的集散、加工与再创造的场所"，办公空间在人们的工作和生活中扮演着不可或缺的角色。近几年来，随着我国社会经济的快速发展以及城市化进程的不断加快，作为城市人居活动"第二场所"的城市办公区的格局发生了较为显著的变化。传统封闭型的办公区逐步走向公共与开放，严肃单一的办公环境逐步走向亲和与多元。

行政办公是城市办公的一大类型，无论是政府机关还是企事业单位，其行政办公区域的环境景观也趋于开放和半开放，并由独享趋于共享，由单一功能逐渐转为功能的多元复合。行政办公区空间景观是城市公共环境的重要组成部分，是城市生产、生活与生态空间的有机结合，展现城市的风貌与文化。景观语言的创新表达目标是改善行政办公建筑冷漠的外部形象、消除高层办公建筑的巨大体量给人带来的压抑感，让环境更亲切宜人，使人感受到政府办公场所的亲民气质。

Throughout history, from the agricultural era to the industrial era, and now the knowledge economy, the acquisition of knowledge and the exchange of information has led to constant continuous improvements in people's lives. As "a place for information distribution, processing and reinvention", office space plays an indispensable role in people's work and everyday lives. In recent years, with the rapid development of China's social and economic development and urbanization, the layout of urban office areas as an alternative venue for the urban activities of residents has undergone dramatic changes. Traditionally closed-off office areas are gradually becoming more public and open. As a result, the previously somber and one-dimensional environment of office buildings has gradually become more approachable and diversified.

Administrative offices are one of the major types of urban offices. Whether a government agency or private enterprise, the layout of these office areas tend to be open or semi-open, and generally tend to prioritize shared spaces over exclusive use, which has gradually turned these previously one-dimensional spaces into multi-functional environments. The landscapes of administrative office spaces are an important part of urban public space. They represent an organic synthesis of urban productivity, urban lifestyles, and ecological space, and reflect the unique characteristics and culture of a city. The innovative expression of landscape design aims to improve the indifferent external image of administrative office buildings and eliminate the oppressive feeling wrought by dense high-rise office buildings. It strives to create a warm and approachable environment, with a more welcoming temperament.

2015年12月，中国城市建设研究院无界景观工作室联合北京市建筑设计研究院、中国建筑设计研究院、北京市弘都城市规划建筑设计院承接了北京城市副中心行政办公区先行启动区景观设计项目，无界景观工作室与北京市建筑设计研究院合作承担了副中心行政办公区先行启动区约70%的景观设计工作。

In December 2015, View Unlimited Landscape Architecture Studio China Academy of Urban Planning and Design, together with the Beijing Institute of Architectural Design, China Architecture Design and Research Group, and Beijing Homedale Institute of Urban Planning & Architectural Design, undertook the landscape design project for the Beijing Municipal Administrative Center Pilot Start-up Area, of which roughly 70% of the landscape design was carried out by View Unlimited and the Beijing Institute of Architectural Design.

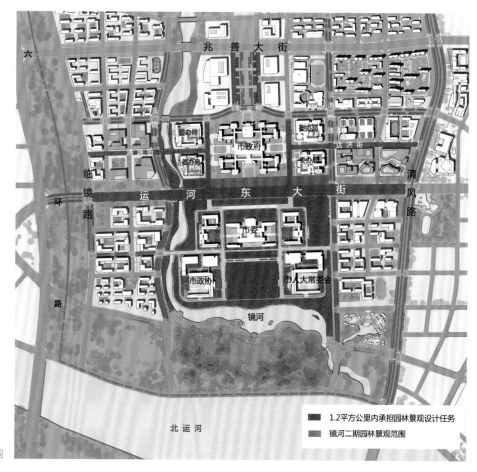

设计范围（先行启动区1.2平方公里）与镜河二期范围

场地建设前原貌

工作伊始，无界景观设计团队在北京市建筑设计研究院朱小地院长主持的北京城市副中心行政办公区总体城市设计指导下，首先对副中心行政办公区总体景观规划做出了方案研究，提出"营造宜人环境、传承北京气质、引领智慧生活"的总体规划目标，为副中心行政办公区先行启动区环境景观设计以及林荫路网、镜河河道景观设计奠定了基础。

"营造宜人环境"——坚持景观规划与行政办公区主体办公建筑统筹协调，从传承中华传统自然观、价值观的理念出发，承袭"可行、可赏、可游、可居"的中国千年城市山水文化理想，营造山水格局，提高环境宜居指数，使建筑与自然相互辉映，使"风景"融入日常生活当中。

"传承北京气质"——中华园林之美在于其浸染了五千年华夏文化的内蕴，承载了中华民族的精神。统筹考虑景观营造与文化内涵，致力于传承大气、包容、自然、大雅的北京气质，在景观营造中体现人文理想。

Initially, View Unlimited's landscape design team, under the guidance of the Urban Planning of Beijing Municipal Administrative Center, chaired by President Zhu Xiaodi of the Beijing Architectural Design and Research Institute, carried out a study of the Beijing Municipal Administrative Center, after which the general planning objectives of "creating an inviting environment, preserving Beijing's heritage, and promoting smart living" were proposed. This laid the foundation for the environmental landscape design to be implemented in the Beijing Municipal Administrative Center Pilot Start-up Area, as well as the landscape design of the green network and Jing River.

"Creating an inviting environment", namely, adhering to the integration of the project's landscape planning and administrative center's main office building, is deeply rooted in the passing down of the traditional Chinese understanding of nature and its values. Out of this, the idea expanded to the country's venerable urban landscapes, which embody the cultural ideals of "accessibility, appreciation, leisure, and comfort", so as to create a landscape that tangibly improves the livability of the surrounding environment and allows architecture and nature to reflect one another by incorporating landscapes into daily life.

"Preserving Beijing's heritage" alludes to the beauty of Chinese gardens, which are steeped in 5,000 years of Chinese culture and symbolize the spirit of the Chinese nation. Considering the intimate relationship between landscapes and cultural connotations, we endeavored to impart Beijing's characteristics of openness, tolerance, nature, and refinement, while reflecting humanistic ideals in the creation of the landscape.

借鉴传统筑山理水的手法，营造山水格局，
传承中国传统"天人合一"的价值观与自然观，
整体形成三河四带的山水关系，使风景融入日常生活。

三河：北运河、运潮减河、镜河。
四带：镜河河谷绿带、六环绿带、北运河及守望林绿带、运潮减河及古城遗址绿带。

北京城市副中心行政办公区山水骨架

"引领智慧生活"——坚持景观设计的早期介入，注重跨领域、多专业合作，促进城市生产、生活、生态空间的互联互通，节约高效地利用土地与资源；综合应用"海绵城市""智慧园林""节约型园林""低碳园林""可再生能源"等先进理念和技术，广泛应用先进材料和工艺，智慧解决城市问题。

"Promoting smart living" advocates for the integration of landscape design in the earliest stages, by focusing on cross-discipline and multi-disciplinary collaboration, promoting the interconnection of urban productivity, everyday life, and ecological space, achieving the conservation and efficient use of land and resources through the wide-ranging application of advanced concepts and technologies such as sponge cities, smart gardens, eco-friendly gardens, low-carbon gardens, renewable energies, etc., and extensively applying advanced materials and technologies to provide smart solutions to urban challenges.

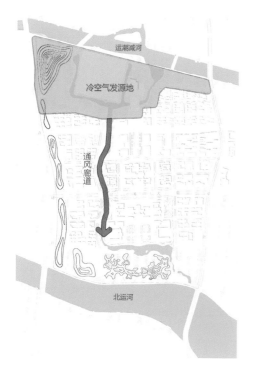  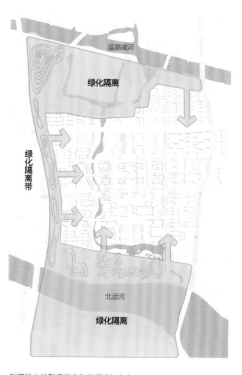

景观布局结构：两带
山水骨架　山水关系

连通水系，发挥滞洪功能，构建通风廊道，
区域水系梳理结合"海绵城市"理念，
拓宽水域加强蓄滞功能，
水系可作为城市片区的通风廊道，促进空气流动。

挖湖堆山，土方平衡，
结合水系开挖塑造地形，区域西北部设置微地形，
六环隔离带抬高地形阻隔六环路噪声。

利用堆山地形促进空气微循环与自净，
利用堆山地形增加风与植物的接触面，
通过植被的吸收和吸附过滤功能，提升片区空气自净能力。

北京城市副中心行政办公区山水骨架分析

02 北京城市副中心行政办公区先行启动区环境景观设计
Landscape Design of the Beijing Municipal Administrative Center Pilot Start-up Area

- **项目地点：** 中国 北京市通州区
- **项目规模：** 1.2 平方公里
- **设计时间：** 2016—2018 年

- **Project location:** Beijing, China
- **Project scale:** 1.2 square kilometer
- **Design period:** 2016-2018

北京城市副中心行政办公区先行启动区位于长安街的东延长线上，运潮减河与北运河之间，是副中心行政办公区的核心区域，总占地面积约1.2平方公里。先行启动区的园林绿化与景观设计严格贯彻落实创新、协调、绿色、开放、共享五大发展理念，贯彻落实中央提出的构建蓝绿交织、清新明亮、水城共融、多组团集约紧凑发展的生态城市布局要求，紧紧围绕副中心战略定位和发展目标，发挥先导与示范作用。

The Beijing Municipal Administrative Center is located in the east extension of Chang'an Road, between the Yunchaojian River and Bei Waterway, and is the core area of the sub-center's administrative office district, covering a total area of approximately 1.2 square kilometers. The greenification and landscaping strictly implements the five major development concepts of innovation, coordination, environmental protection, openness, and sharing, so as to realize the central government's requirements for creating ecological urban layouts that closely integrate nature with revitalizing elements, water/city integration, and intensive and concentrated development consisting of multiple clusters. Thus, by carefully focusing on the strategic positioning and developmental goals of the sub-center, we were able to play a pioneering and demonstrative role.

(1) 让建筑融于树林之中，营造舒适宜人的森林行政办公区

北京城，本来就是一个只见树木不见屋顶的绿色都会。——郁达夫《北平的四季》

设计团队以"大林汇智，经纬交织"为总体设计理念，提出"大树林、风景环、景观轴、园中园、绿荫厅、健身道、林荫路、海绵网"的特色景观结构，从面、环、线、点、网多种维度，将场地内行政办公建筑群植入绿色环境中，塑造端庄大气却又不失灵动的景观氛围，形成人与自然相融的舒适环境，使"风景"融入日常生活之中。

(1) Blending the Building into the Forest to Create a Comfortable and Inviting Administrative Office Area

In the past, the city of Beijing was a green metropolis of trees as far as the eye could see.—— Yu Dafu, *Four Seasons in Beiping*

The design team employed an overarching concept of "the interweaving of nature and wisdom" and proposed installing a landscape characterized by roaming woods, a scenic belt, scenic axis, and a garden within a garden, in addition to shade structures, fitness trails, shaded paths, and a sponge-like network that incorporated the administrative office buildings into the surrounding green environment based on a variety of factors, such as surfaces, rings, routes, sites, and networks to foster am atmosphere with a sense of poise and dynamism, and forming a comfortable environment where people and nature blend together harmoniously, by integrating landscapes into daily life.

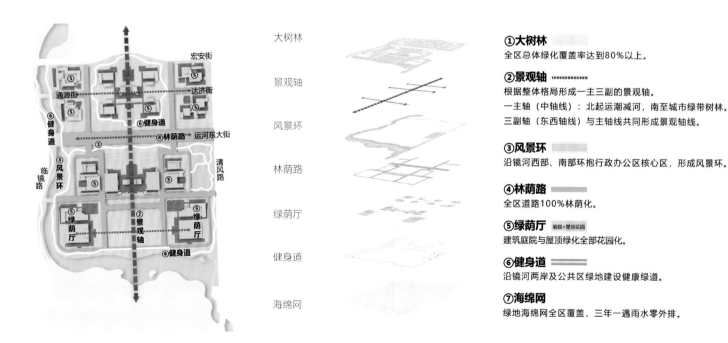

北京城市副中心行政办公区先行启动区景观结构图

森林中的行政办公区（丁洪兴摄影）

丰字沟原貌,后改名为"镜河"

风景环环绕绕下的副中心先行启动区(丁洪兴摄影)

· 大树林

"大树林"特色景观设计遵循低碳节约型园林设计理念，通过绿环、绿网、绿轴、绿庭、绿带的系统设计，在空间上满足人们交流、休憩、健身的需求；结合微地形改造，在营造林中办公环境的同时，因地制宜地营造舒适小气候与多样生境，植被主要选用本地植物，打造生态化的园林绿化创新示范区。

· 风景环

环绕在地块西南侧的镜河风景河道❶是"风景环"❷的重要载体，作为润城之水，孕育蓝绿交织、水城共融的滨水空间。

北京城市副中心行政办公区先行启动区大树林景观（耿毅军摄影）

· **The Great Woods**

The Great Woods are characterized by their low-carbon and energy-efficient design, which systematically utilizes green rings, green networks, a green axis, green grottoes, and green belts to meet people's needs for social interaction, leisure, and health and fitness. By applying minor modifications to the existing terrain, an office environment was created in the forest which offers a comfortable micro-climate and diverse habitats that took local conditions into consideration and mainly used indigenous plants to create an ecological garden and demonstration area for green innovation.

· **Scenic Ring**

The scenic Jing River waterway, which surrounds the southwest side is an important element of the site'[3]s scenic ring[4]. As the city's proverbial wellspring, it hearkens back to Beijing's ancient roots, while creating a waterfront area that closely integrates nature and water/city integration.

风景环
从使用人群需求出发，提升宜居指数，将风景融入日常生活，
镜河河道全长约4千米，水域宽23~156米，
起调水位18米，最高水位20米。

风景环环绕北京城市副中心　　　　　　　　　　　　　　　　　　　　　镜河风景环是城市副中心的润城之水

❶ 详见62页"北京城市副中心行政办公区镜河道景观专项设计"。
❷ 详见40页资料链接1。
❸ For further details, see page 62 'Landscape Design of the Beijing Municipal Administrative Center Jing River Area'.
❹ For further details, view page 40 Link 1.

水城共融、蓝绿交织的滨水城市景观

傍晚从镜河西岸向东岸望

"风景环"镜河河道春景

秋季的景观河谷

· 景观轴

由办公建筑与外部空间共同构成经纬交织的"景观轴",南北为主,东西为辅,形成中正大气的礼仪性空间序列;市委南广场❶是景观主轴线的南端起点和重要节点,与千年城市守望林隔镜河相望。

· Landscape Axis

The office buildings and external spaces form a landscape axis, with the north and south as the main points and the east and west as the secondary points, which creates a spatial progression that is equitable and inclusive. The south plaza❷ is the starting point and a vital node for the main landscape axis, which overlooks the Jing River together with the Forest of Guidance.

❶ 详见42页链接资料2、46页链接资料3。
❷ For futher details, view page 42 Link 2, page 46 Link 3.

上图/先行启动区南北向景观轴(耿毅军摄影)
下图/从水岸边看先行启动区南北向景观轴(丁洪兴摄影)

景中有景，园中有园
富有意境的文化景观

由区内园路串联

功能复合
开放共享的户外交流空间

雨水收集与利用
智慧园林设施的展示空间

"坚固而不笨拙，
华丽而不俗气，
灵巧而不单薄，
轩敞而不空旷，
幽静而不阴暗。"
—— 朱家溍

园中园愿景

· 园中园

"园中园"体现中国传统集锦式园林手法与现代景观设计的融合，使公共空间"园中有园、景中有景"。尽可能降低建筑的体量感，减弱建筑对环境的影响，让自然渗透建筑，通过景观设计手法将室外的游径、观景休闲平台，与植被、景观水池组成不断变化的空间片段，形成不同视觉场景的景点与观景点。用区内园路串联富有意境的文化景观，构建功能复合、开放共享的户外交流空间，同时也是雨水收集与利用、智慧园林设施的展示空间。

· Garden Within a Garden

The concept of "a garden within a garden" reflects the integration of traditional Chinese gardening techniques and modern landscape design, with the goal of creating public spaces that are "a garden within a garden and a landscape within a landscape". To the greatest extent possible, the team aimed to reduce the sense of volume and environmental impact of the buildings, and let nature permeate and flow into the architecture. By applying landscape design techniques, outdoor trails and viewing platforms, together with vegetation and landscape pools, ever-changing spatial segments were created that form attractions and viewing areas offering a variety of stunning scenes. The area's garden paths are used to connect landscapes with rich cultural connotations and create an open and shared outdoor space with a variety of integrated functions and a demonstration area for rainwater collection and utilization and intelligent garden facilities.

· 绿荫厅

环境舒适、尺度宜人的"绿荫厅",亦是横纵交织的"城市立体森林",包括花园式的建筑庭院和功能复合的屋顶花园,满足会客、交流、健身等多种活动功能。不同尺度的非正式交往空间有序交织,打造功能复合的户外"客厅",形成不同层次的、人与自然交流的场所,满足不同使用人群提神醒脑、缓解压力等不同的需求。庭院景观设计是建筑的延续和变奏,强调院落感的体验,在有序的组织下增添一分灵动活泼;可上人的屋顶花园是建筑室内办公的延伸,通过"室内化"的设计手法,采用复层种植设计营造舒适、精致的空间环境。

· Shade Structures

Shade structures offer a comfortable environment and "vertical urban forest", with interweaving horizontal and vertical structures, including garden-style courtyards and functional rooftop gardens, which support a variety of activities such as meetings, social events, and health and fitness. Informal social spaces of various scales are organized in a set pattern, forming staggered levels for communication and exchange between people and nature, which meet the needs of individuals with regards to refreshing the mind and relieving stress, while creating a functional and integrated outdoor "parlor". The courtyard landscape design is both a continuation and variation of the building's architecture, which emphasizes the experience of the courtyard lifestyle, while introducing greater flexibility and dynamism under an organized structure. The roof garden, which can be used by people, is an extension of the building's indoor offices, and creates a comfortable and exquisite spatial environment by implementing the design approach of "indoorization" and the use of integrated planting.

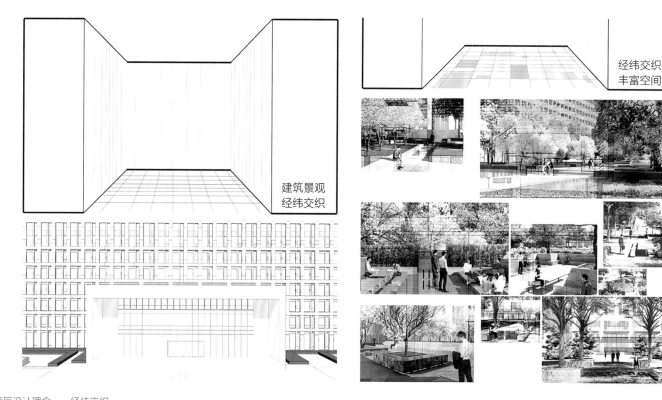

绿荫厅设计理念——经纬交织
庭院景观设计与建筑紧密结合,是建筑的延续和变奏,在有序的组织下更加灵动活泼。
不同尺度的非正式交往空间有序交织,林荫庭院、四时景观、提神醒脑、缓解压力。

北京城市副中心绿荫厅"经纬交织"的设计理念

办公区内横纵交织的"城市立体森林"(耿毅军摄影)

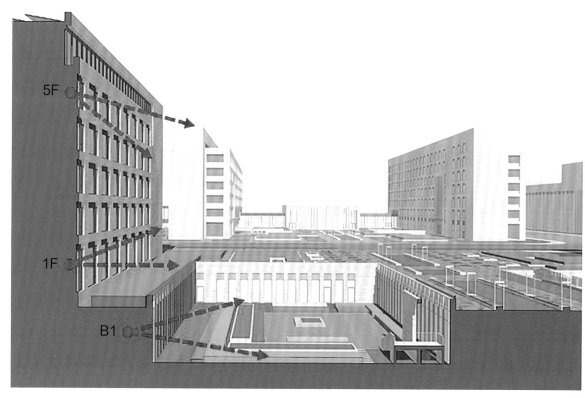

林荫庭院为不同高度的视线提供一片绿意,缓解眼部疲劳,舒缓心情

下三图/对办公区内的景观精细化设计使办公楼窗窗有景

· 健身道、林荫路、海绵网

构建一体化设计的"林荫路"网和环网结合的立体绿道、"健身道"系统，串联弹性场地空间，并叠加"海绵网"功能，使路网与场地兼具雨水调蓄、补给地下水和乡土生境保护等多重功能❶。

· Fitness Trails, Shaded Paths, and a Sponge-like Network

The integrated design of shaded paths, a vertical greenway combined with a ring network, and fitness trails connects the site's flexible spatial arrangement and superimposes a sponge-like network, so that the network of paths and the site itself can support multiple functions, such as rainwater storage, groundwater restoration, and conservation of the native habitat❷.

林荫路网愿景

❶ 详见48页"北京城市副中心行政办公区先行启动区林荫路网专项设计"。
❷ For further details, see page 48 'Design of the Beijing Municipal Administrative Center Pilot Start-up Area Shaded Path Network'.

镜河河谷健身道串联城市绿道网，形成了移步换景的绿道

林荫健身道设置了许多看得见风景的休憩之处

（2）在环境中传承北京气质，塑造开放共享的弹性公共空间

北京的城市气质，首在大气、包容。

北京历元明清三代至今，海纳百川，五方杂处。贵族文化、文人文化与胡同市民文化兼容。不同文化在并存共生中，不但不违和，还相得益彰。贵族文化大雅近俗，胡同市民文化俗中有雅，源远流长的文人文化则为城市赋予了浓厚的文化气息。改革开放后，北京的兼容并蓄又体现在对异质文化的吸纳。传统与现代衔接，中西交融，不仅不违和，而且在变动中保存了城市特有的格调、气象。

"大气""包容"体现于北京的空间形态，是皇家园林的开敞，布局疏密有致；是胡同的规整而有变化，不逼仄局促。设计团队在北京城市行政副中心的景观设计中，既借鉴皇家园林的恢宏阔大，又融入胡同文化的平易亲和，兼顾格局、气势与人们日常的使用需求。郁达夫所赞许的"具城市之外形，而又富有乡村的景象之田园都市"（《住所的话》），有可能实现在这样的景观营造中。

(2) Inheriting Beijing's Characteristics and Shaping a Resilient Public Space that is Open and Shared

Beijing's urban character is defined first and foremost by its openness and tolerance.

From the Yuan, Ming and Qing dynasties to its modern-day incarnation, the city of Beijing has accommodated a diverse range of residents hailing from all across China. The aristocratic and literati cultures have historically been compatible with the civic culture of the hutongs. Rather than clashing, the coexistence of these different cultures are complementary. The aristocratic culture is elegant and almost licentious, while the civic culture of the hutongs has a folksy charm. The long-standing literati culture ferments the strong cultural flavor of the city. After the period of "reform and opening up", Beijing's inclusiveness began to be reflected in its absorption of diverse and foreign cultures. As tradition and modernity have converged in the East meeting the West, rather than creating incongruity, the unique style and atmosphere of the city has been preserved in the midst of these changes.

This "openness" and "tolerance" is reflected in the city's spatial form, with open royal gardens and a layout that is simultaneously sparse and dense, and also in the hutongs, which are uniform and varied, but neither cramped nor confined. For the landscape design of the Beijing Municipal Administrative Center, the design team drew on the grandeur of the royal gardens, while incorporating the friendly, down to earth character of the hutong culture, with a harmonious layout and atmosphere that closely reflects people's daily needs.

借鉴皇家园林的恢宏阔大，形成"大气""包容"的整体环境氛围

传承北京气质，塑造开放共享的公共空间

镜河河道滨水城市阳台

"田园都市"的精髓在于人与自然的和谐。风景不仅供观赏,更遍布于人的生活空间,是人们日常环境的一部分。设计团队致力于在景观营造中体现人文理想,以富于文化内涵的设施,使人文风物浑然一体,对生活其中者产生潜移默化的影响。

老舍在其作品中形容北京老派市民的教养、风度为"自然、大雅"(《四世同堂》)。这岂不也应当是北京作为城市应有的风度?

The essence of an "idyllic city" exists in the harmony between humans and nature. The landscape is not only a spectacle to be viewed, but permeates throughout people's living space and is a part of their daily environment. In creating this landscape, the design team expressed a commitment to embodying humanistic ideals, with a range of elements that are rich in cultural connotations, so that this heritage may be integrated into the landscape's design and exert a subtle influence on its residents.

In his work Four Generations Under One Roof, Lao She described the upbringing and manners of the local citizens of Beijing as "natural and elegant". Should this not also be reflected in Beijing's character as a city?

上图/"田园都市""城市森林"里的行政办公区(丁洪兴摄影)

下图/造型油松与庄重大气的铺装展现办公区的环境气质(耿毅军摄影)

初春本土植物衬托出庄重大气的办公区形象

设计团队以新文化运动后文人笔下的意境为灵感，更以熟悉的北京传统院落景观为蓝本，以古今交汇、传统与现代不同意象的组合与变化为特色，巧妙运用传统铺装和特色构筑物，结合使用功能与建造需要，呈现北京的非物质文化遗产，承载传统文化，传承大气、包容、自然、大雅的北京气质。

The design team was inspired by the ideals of the literati after the New Culture Movement, and by the familiar traditional courtyard landscapes of Beijing, which feature the intersection of ancient and modern elements, combinations and variations of different imagery representing tradition and modernity, the clever use of traditional paving and distinctive structures, the integration of functions with construction needs, and the presentation of Beijing's intangible cultural heritage, which conveys its rich traditional culture and unique character of openness, tolerance, nature, and refinement.

秋季风景充分展现了"田园都市"中人与自然的和谐景观

结合绿地设置富有文化内涵的公共设施,塑造弹性开放的公共空间,复合教育、科普、健身、公共艺术❶等功能,以润物细无声的方式将北京气质赋予场地之中,在守护"活力基因"的过程中,探寻景观影响人的有效途径,缓解人的生存压力,提升场地活力和吸引力。

By combining green space with public facilities steeped in cultural connotation, shaping flexible and open public space, incorporating education, science, health and fitness, public❷ art and other functions into the site, and embodying the character of the city, we seamlessly integrated the landscape with Beijing's historical heritage and customs. On the premise of ensuring the site's dynamism, we used the landscape to explore effective ways to have an impact on people, relieve the pressures of human existence, and enhance the site's vitality and aesthetic appeal.

台阶与看台结合,提供弹性休息空间

❶ 详见119页资料链接5、129页资料链接6。
❷ For further details, view page 119 Link 5、page 129 Link 6.

（3）智慧解决城市问题，建设统筹协调的创新示范园林

设计团队所进行的是对紧凑高效的土地使用模式的探索，有助于保护开敞空间，减少空气污染，具有复合功能和适宜步行的特点，减少人们对汽车的依赖和开支，使基础设施得到充分利用，最终将自然环境与人工区域结合成一个可持续的、有机整体的、功能化和艺术化的步行网络，形成使用者认可的、具有归属感和安全感的充满人情味的生态园区，并促进使用者形成对保持园区活力、促进社区发展与进步所必需的责任感。

(3) Smart Solutions to Urban Problems and Integrated and Coordinated Building Methods for Innovative Demonstration Gardens

The design team explored concentrated and efficient models for land use that help preserve open space and reduce air pollution, with a focus on multi-functional and pedestrian areas which reduce people's dependence on cars and other vehicle-related expenditures, while making full use of the available infrastructure to ultimately combine the surrounding natural environment and man-made areas into a sustainable, organic, functional and aesthetically pleasing pedestrian network that forms a humane eco-park and fosters a sense of belonging and security, while promoting a sense of social responsibility necessary to maintain the health of the park and promote community development and advancement.

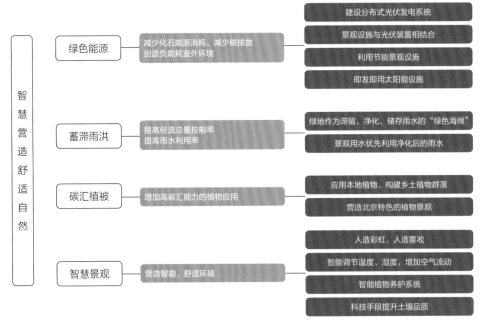

一体化营建的生态景观系统

针对北京市城市副中心局部小气候进行改善与优化，关注使用者在城市活动空间中对气候环境的人体感受。研究生产微气候系统运用于城市副中心镜河河道，结合场地设置微雾风扇，通过感应装置感应实时温度、湿度，开启风扇与雾效，调节场地局部空气湿度及温度，减少浮尘等，形成舒适宜人的小气候

设计团队坚持在前期就介入行政办公区总体规划与策划，广泛应用先进材料和工艺，通过多专业技术手段的创新一体化应用，统筹协调交通、能源、生态环保、水利、电信等相关部门进行跨领域协作，以共享模式整合多种公共空间，使绿化景观与整体环境和谐统一、资源和环境得以高效利用，最大限度地实现绿色环保、低碳节能、舒适宜居，发挥生态园林的创新示范作用。

The design team focused on playing an active role in the planning of the Beijing Municipal Administrative Center in the earliest stages through the broad application of advanced materials and techniques, focusing on cross-disciplinary and multi-disciplinary collaboration across relevant departments, such as transportation, energy, ecology and environmental protection, water conservancy, and telecommunications, and utilizing the innovative integration of multi-disciplinary techniques, so as to integrate a variety of public spaces based on a shared model, harmonize the relationship between the landscape and overall environment, make efficient use of available resources, maximize environmental protection, energy conservation, and the overall comfort of the area to the greatest extent possible, and demonstrate the innovative role of ecological gardens.

利用景观构筑物设置了薄膜光伏板，储能蓄电，为照明、风扇提供电力。按北京市年平均每日有效日照4.2小时计算，每日发电约14千瓦时，年发电约5100千瓦时。廊架内设置智能感应装置，通过感应照度、温度、湿度，实时控制照明、风扇等，调节休憩场地的湿度及温度，营造舒适的休憩环境。同时廊架设有红外感应装置，有人时才会开启照明与风扇，避免资源浪费

在镜河景观的设计过程中，设计团队提出了将道路附属绿地与公园绿地一体化统筹设计的布局模式，将市政道路人行道与公园园路合并，减少硬质铺装，拓宽公园绿地。通过对奥林匹克森林公园、昆玉河河岸作为主要研究对象进行大量的数据采集与统计，以不同的测量仪器、软件及不同的测量方法进行实际环境数据的测量统计、分析得出实际数据，再与ENVI-MET动态模拟数值相互比较、验证拟合。同时建立副中心镜河河道不同的绿地与道路格局小气候动态模型，对比奥林匹克森林公园、昆玉河河岸几种绿地格局类型的小气候环境特征验证模型的有效性。主要模型指标包含温度、湿度、太阳辐射、风速风向、噪声等。在此基础上，深入分析镜河河岸公共空间的高宽比、绿化覆盖率、铺装材料、高差等与小气候模型直接的关联性，最终提出影响副中心公共空间小气候环境的可能因素及改善策略。

设计整体统筹城市慢行路网、绿道网、道路海绵网，对道路附属绿地与公园绿地进行一体化设计，减少硬质铺装，拓宽公园绿地；调整非机动车道位置，拓宽机非隔离带，进一步降低机动车车道对骑行的影响；结合集雨型绿地建设，落实海绵城市理念。

During the landscape design process for the Jing River, the design team proposed a model that integrated green streets and park spaces into the design by merging sidewalks with garden pathways, reducing the prevalence of hard paving and expanding the available park space. By referencing the Olympic Forest Park and Kunyu River riverbank, the team amassed a vast amount of information, and applied various instruments, software, and survey methods to accrue accurate environmental data and statistics. Then, the team performed a comprehensive analysis, which was compared with values from ENVI-met dynamic simulations to validate the survey data with corresponding models. At the same time, the team applied a dynamic microclimate model to the various green spaces and pathways that comprise the Jing River. The environmental characteristics of several green space microclimates in the Olympic Forest Park and Kunyu River riverbank were then analyzed and compared with the Jing River's green spaces to verify the validity of these simulations. The main metrics used in the model included temperature, humidity, solar radiation, wind speed and direction, noise, etc. On this basis, a direct correlation between height-width ratio, green coverage, paving materials, and height disparities of public spaces along the Jing River and corresponding microclimate models were analyzed in detail and possible factors and strategies to improve the effects of each individual microclimate environment were proposed.

An urban slow-moving traffic network, greenway network, sponge road network, and green spaces adjacent to the road were integrated into the park's green spaces, thus reducing the prevalence of hard pavement, and expanding the available green space. By reorienting the placement of non-motorized lanes and emphasizing the separation between motorized and non-motorized vehicles, the design team further reduced the impact of motorized lanes on cycling routes and developed rain-collecting green areas to successfully implement the concept of a sponge city.

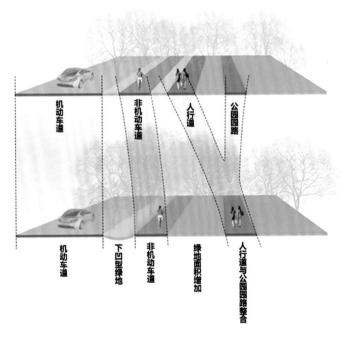

道路附属绿地与公园一体化设计可以高效利用土地空间，提高城市绿量

经过模拟验证，镜河一体化林荫路可有效降低温度约2.1摄氏度；园路能够有效增湿12%；城市路面在完全暴露的情况下，太阳辐射可达1029瓦每平方米，而树荫可以有效降低太阳辐射，河道内林荫健身步道模拟结果为618瓦每平方米；结合有效的竖向变化、绿地宽度及种植对噪声的衰减作用，大部分空间噪声可以衰减5~10分贝；利用植物树叶的滞尘作用，林荫健身步道及林荫开放空间比城市平均PM2.5值低3微克每立方米、平均PM10值低8微克每立方米；河道作为通风廊道拥有更好的空气流通性。

After verifying the simulations, the integrated riverwalk effectively reduced temperatures by approximately 2.1℃ while the garden paths effectively increased humidity by 12%. When fully exposed, solar radiation on the urban road surface reached 1029W/m², however this was effectively mitigated by the shaded areas (simulations of the shaded fitness trail along the river estimated solar radiation of 618W/m²). By integrating effective vertical developments with expanded green space and greenery, the design team were able to attenuate environmental noise, with a noise reduction of 5-10 dB on average and by utilizing the dust retention ability of plants, the shaded fitness trail and shaded open spaces were 3μg/m³ and 8μg/m³ lower than the average PM2.5 and PM10 values in the city respectively, while the river offered more optimal air circulation as a ventilation corridor.

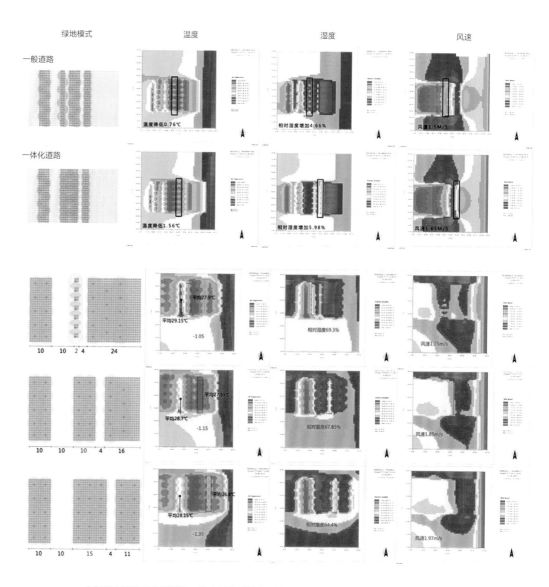

人行道距离机动车道越远，结合植物群落降温效果越明显。相对湿度与温度呈现负相关，湿度高于65%时会抑制人体排汗容易感到闷热。风速的提高，能够降温，提高舒适感

资料链接1
市人大委员会办公区东侧"叠水花溪"景观设计

镜河及市人大委员会办公区东侧"叠水花溪"构成了围绕办公区建筑的"风景环",共同构建了北京城市副中心行政办公区先行启动区的山水骨架。

花溪的驳岸采用了叠石的形式,意在体现中国传统园林"叠山理水"的造园手法,通过大小错落、起伏变化的山石修建护坡,同时形成深邃幽远、蜿蜒生动的水形。花溪的设计不仅丰富了竖向的空间变化,还作为区域"海绵网"的一部分,充分发挥了水体景观的滞洪、增湿、通风、降温、降噪等生态功能,是景观与海绵一体化统筹建设的实践。

花溪建造时,由北方从事叠石掇山近四十年的工匠师傅亲自指导,采用山石韩叠山技艺(南派假山技艺的代表),对从北京近郊及周边城市运来的石头进行挑选,主要选用北京房山地区、河北涞水等地的千层石,因地制宜、随形就势叠石筑岸。景观设计师在此基础上利用不同植物材料进行搭配,打造充满自然野趣、四时变化的水岸景观。

慢行路穿梭于叠水花溪之间

Link 1

Landscape Design of the "Cascading Floral Stream" to the East of the NPC Office Area

The Jing River and cascading floral stream to the east side of the NPC office area form a scenic ring around the buildings in the office area. Together, they create the landscape framework of the Beijing Municipal Administrative Center Pilot Start-up Area.

The revetment along the floral stream adopts a stacked rock formation, with the intention of reflecting elements of traditional Chinese gardening techniques of stacked mountains and flowing water. The slope is shaped by rocks of varying sizes and undulations, so as to form a profound, far-reaching and vivid water feature. The design of the floral stream not only enriches the area's vertical space, but also gives full play to a host of ecological functions related to the water landscape such as flood control, hydration, ventilation, temperature cooling and noise reduction and practically integrates a sponge network into the development of the landscape.

使园林拥有滞洪、增湿、降温等生态功能

以"叠山理水"为造园手法设计的叠水花溪

资料链接2
市委办公区南广场景观设计

在朱小地院长总体设计理念的指导下，市委办公区南广场整体设计延续中轴对称的景观结构，与建筑经纬交织、相融共生，在有序的组织下更显灵动活泼。

中轴线设计为开敞的景观草坪，两侧分布步行广场，南北主轴配以北京市树国槐组成的树阵，步行广场外侧列植常绿乔木油松，东西次轴绿地列植国槐，形成兼顾四季景观的礼仪性轴线。

轴线两侧绿地基本采用对称式布局手法，既符合中国传统的对称与均衡理念，同时满足整体礼仪性的需要。将花园、景观镜面水池和人行步道巧妙结合，水池环绕在花园之中，将各个绿地串联起来，水随路转、绿因水活、曲径通幽、步移景异，处处彰显传统文化气韵，营造出变化丰富的景观空间。

南广场水池沿市委办公区延续而成的中轴线对称分布，市人大委员会和市政协办公区形成的东西轴线与市委办公区中轴线将南广场划分为多个绿地空间，设计不同大小尺度的停留空间与宽窄相异的步道镶嵌至绿地中，繁复求变、乱中有序。相同尺度的树池、木地板、艺术石凳不断重复，表达整体设计的节奏变化与韵律，体现材质、空间的对比与调和，将疏密、大小、主次、虚实、动静、聚散进行协调统一，使整体景观效果达到均衡，做到整体统一，又富有局部变化。

镜面水池为南广场绿地增添了灵动性与互动性，对称布置的镜面水池，使南广场整体的视觉效果达到统一，更具组织性。有水时，水池倒映建筑、天空、植物与人的活动；没水时，水池则成为多功能活动场地，体现功能复合、弹性设计理念。穿插在绿地中的竹木地板，可局部放大停留空间，形成功能复合的交往空间，兼具停留、赏景、非正式交谈等多重功能，达到不同尺度的交往空间有序交织。

南广场旗台是市委和市人大委员会、市政协办公区两轴交汇的中心，周边配以小型广场，国旗台为简约现代的三层白色花岗岩平台，排砖严整，既符合中国传统文化，又与周边建筑风格相协调，礼仪感与秩序感强。

Link 2
Landscape Design of the Municipal Office South Plaza

Under the guidance of President Zhu Xiaodi's design concept, the overall design of the Municipal Office South Plaza perpetuates the central axis' symmetrical landscape, while integrating the surrounding architecture in a unified and symbiotic style and using an orderly structure to instill vibrant and dynamic elements.

The central axis is designed as an open landscaped lawn, with pedestrian plazas on both sides. The north-south main axis features lines of *Sophora japonica* native to Beijing, with evergreen *Pinus tabuliformis* planted along the edges of the pedestrian plaza. More *Sophora japonica* have been planted in the green spaces along the east-west sub-axis, forming a ceremonial axis that takes all four seasons into consideration.

The green spaces on both sides of the axis adopt a symmetrical layout, which reflects the traditional Chinese concept of symmetry and balance, while meeting all of the feature's ceremonial needs. The gardens, landscaped mirror pools and pedestrian walkways have been masterfully combined. The pools, surrounded by gardens, link together various green spaces, while their waters flow along the various paths, infusing the surrounding greenery with a sense of vitality. The winding paths offer breathtaking views, with different scenic features at every step that highlight the charm of traditional Chinese culture and create rich and varied landscapes.

The South Plaza's pools have been symmetrically formed by the central axis, which is a continuation of the municipal office area, while the east-west axis formed by the NPC, CPPCC and aforementioned municipal office areas divide the South Plaza into several green spaces. By employing a design that utilizes resting spaces and walkways of various scales, which are integrated into the green spaces, a complex and evolving landscape has been achieved that reflects a sense of order in chaos. A corresponding scale has been applied to the trees, ponds, wooden planks, and artisanal stone benches repeated throughout the landscape, so as to express a cadence of change in the context of the rhythm of the overall design and reflect the contrast and harmony of materials and spaces. Concepts such as density and size, major and minor elements, real and illusory, static vs. dynamic, and clustering and dispersion have been harmonized, so that the overall effect of the landscape can achieve a balance and unity, while emphasizing subtle changes.

The mirror pools infuse a dynamism and interactivity into the South Plaza's green spaces, while the symmetrical arrangement of the mirror pools lends a unified and structured overall visual effect to the South Plaza. When the pools are filled, their waters reflect the buildings, sky, plants, and human activity; when they are empty, the pools are transformed into multifunctional sites which reflect the design team's concept of an integrated and flexible design. The bamboo and wood planks interspersed throughout the green spaces offer the opportunity to extend resting spaces, while forming integrated and

南广场的种植设计突出轴线，由玉兰、海棠等组成春花组团，由油松、白皮松组成常绿组团，点缀椿树、白蜡、元宝枫等点景树，使开敞空间与密闭空间完美结合，营造"三季有彩、四季常绿"，令人感到熟悉又欣欣向荣的北京庭院式植物景观，让使用者和参观者感到亲切的北京气韵。南广场的骨干树——近百株国槐，是运河西大街改造升级移除的老行道树移栽至此，既满足了即时的绿化景观效果，也使得更新换代的老行道树物尽其用，体现了节约型园林的理念。

interactive spaces that support various functions such as resting, scenic viewing, and informal conversations, so as to achieve a structured integration of interactive spaces on a broad scale.

The flag platform in the South Plaza forms the center of the intersection of the two axes of the municipal offices and NPC and CPPCC offices and features a small surrounding plaza. The flag platform is a simple, modern three-tier white granite platform, which not only conforms to the aesthetics of traditional Chinese culture but reconciles the surrounding architectural style by embodying a powerful sense of etiquette and order in the feature's meticulous brickwork.

In addition to highlighting the axis, the design of the South Plaza's greening focuses on *Magnolia x soulangeana* and *Malus x micromalus* to form a cluster of spring flowers, supplemented with *Pinus tabuliformis* and *Pinus bungeana*, which form an evergreen group that is dotted with *Toona sinensis*, *Fraxinus chinensis*, *Acer truncatum* and other scenic trees, so that open and closed spaces are artfully combined to create a familiar and flourishing courtyard-style gardenscape that embodies the "three seasons of vibrant colors and four seasons of evergreen" that allows occupants and visitors alike to resonate with the familiar and intimate feeling of Beijing's atmosphere. Nearly 100 *Sophora japonica* which form the backbone of the South Plaza were transplanted during the renovation of Yunhe West Street. These aged trees not only provide an immediate and effective greening solution but imbue a sense of renewal through the concept of garden conservation.

施工中的南广场两侧绿地（耿毅军摄影）

南广场两侧景观绿地（耿毅军摄影）

资料链接3

《只争朝夕》南广场地面艺术装置作品

《只争朝夕》地面艺术装置作品位于市委办公区南广场，南临镜河水系，与千年守望林隔岸相望，是景观主轴线的重要端点。

该作品由著名艺术家黄海涛领衔创作，无界景观设计师与北方石匠共同参与，占地面积约1400平方米。

汉白玉与青石是北方最能象征城市营造的物质意向。作品运用当代艺术手法将具有百年以上历史、承载着传统手作痕迹的汉白玉、青石（20块老汉白玉、51块老青石）与当代机械加工的花岗岩相互编织、榫接，形成过去—现在—未来的意向（其中有三对老石榫卯通过新的白麻花岗岩构件被重新榫接在一起）。

作品将不同时代的石材组合在一起，形成抽象与写意的视觉效果，以此表达对于千年营城智慧的感叹与共鸣，唤醒人们对于历史文化脉络的追忆，激活人们对于自身文化基因的思考。

此作品的汉白玉、青石主要取自老北京四合院的门槛、台阶石等，这些具有古老手作传统痕迹的老石材，有着经时光沉淀而形成的润泽亲切的肌理，其朴素与深沉可以起到强化室外大空间深厚感的作用。

作品中朴素沉稳的老石材，每日晨昏会随阳光的变化展

Link 3

"Seize Every Minute" South Plaza Art Installation

The "Seize Every Minute" ground-level art installation, which is located in the South Plaza of the Beijing Municipal Administrative Center to the south of the Jing River water system and across from the Forest of Guidance, functions as an important endpoint of the sub-center's axis.

The work was led by the famous artist Huang Haitao, with the participation of View Unlimited Landscape Architecture Design and Northern Stonemasons, and covers an area of approximately 1,400 square meters.

Chinese white jade and limestone, rich with cultural connotation, aptly symbolize the development of cities in northern China. The work uses contemporary techniques to weave and intertwine Chinese white jade and limestone, which bear the traces of traditional handwork (20 pieces of aged Chinese white jade and 51 pieces of aged limestone more than 100 years old), with contemporary machined granite to form a past-present-future motif (to support the piece, three pairs of aged stone mortise and tenons have been joined together with new white hemp granite components).

The work combines stones from different eras to form an abstract and realistic effect, thus expressing a sense of resonance and lamentation for the wisdom of China's venerable Yingcheng (ancient city planning), while at the same time awakening the memory of China's historical and cultural lineage and inspiring greater reflection on our own cultural traditions.

The Chinese white jade and limestone in this work have been procured mainly from the threshold and stepping stones of old Beijing courtyards. These aged stones, with traces of the heritage of ancient craftsmanship have a smooth and familiar texture formed by the precipitation of time. Their simplicity and rich history play an important role in

艺术家黄海涛老师现场指导施工

现出不同的色彩效果：润泽的表面会在朝曦与晚霞的映衬下反射出由淡黄、橘红到玫瑰红色的光芒，与十二棵松树、宽阔的河面一起组成壮丽灿烂、稳重大气的景观。

作品名称源自毛主席诗词："多少事，从来急，天地转，光阴迫。一万年太久，只争朝夕。"

地面艺术装置作品的施工建设要感谢曲阳石雕艺术家刘文亮先生的无私奉献，他为项目捐助私人藏品，保证了最终的建设效果。

strengthening the sense of profundity in the large outdoor space.

The aged stones featured in the work are rustic and sedate, and reveal different colors as the sunlight changes throughout the day. Its lustrous surface reflects the light of the morning and evening sun, from yellow, to orange and finally a rosy red hue, thus forming a magnificent and splendid landscape together with the twelve pine trees and majestic river.

The title of the work is inspired by the poem written by Chairman Mao: "The world rolls on, time presses. Ten thousand years are too long, seize the day, seize the hour!"

地面雕塑《只争朝夕》（耿毅军摄影）

文化艺术与景观结合体现北京气质与工匠精神（耿毅军摄影）

市委南广场地面雕塑《只争朝夕》古今辉映，保留的老树延续了场地的记忆与脉络

03 北京城市副中心行政办公区先行启动区林荫路网专项设计
Design of the Beijing Municipal Administrative Center Pilot Start-up Area Shaded Path Network

项目地点： 中国 北京市通州区

项目规模： 道路总长度约 7.6 公里

设计时间： 2016—2018 年

Project location: Beijing, China

Project scale: 7.6 kilometer

Design period: 2016-2018

先行启动区的一条景观轴线贯穿南北，三条东西向轴线相互呼应，镜河河道与建筑公共绿地组成环抱办公核心区的生态围合，整体构建全区山水结构与自然式的园林景观。场地内林荫路网包含运河东大街等五条市政路及区内道路，总长度约 7.6 公里（包含市政道路 2.2 公里）。

设计团队提出将林荫路网、健身路网、海绵网统筹安排、一体化设计，以"一体化街道"构建先行启动区的绿色骨架。综合应用"海绵城市""智慧园林""节约型园林""低碳园林"等先进理念和技术，结合道路等级进行环境优化调整，注重安全、舒适的绿色出行体验，创建 100% 步行林荫路，落实副中心森林行政办公区定位，打造生态化园林绿化创新示范区。

景观布局结构：以"多线"绿道与城市绿道系统衔接。

利用镜河和六环绿带营建绿道，构建运潮减河绿道和北运河绿道之间的联系。

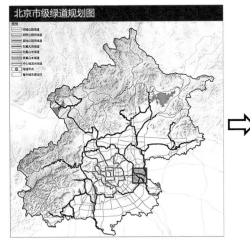

规划两纵一环绿道线路衔接北京市级绿道，联系周边公园绿地，完善休闲健身网络。
（资料来源：《北京市级绿道建设总体方案（2013—2017 年）》）

北京城市副中心行政办公区先行启动区绿道系统与北京市绿道网衔接

A landscape axis runs from north to south, and three east-west axes have been designed to correspond to each other. Together with the Jing River and public green space around the buildings, an ecological enclosure has been created around the core office area which achieves a natural garden landscape. The shaded path network contains five municipal roads, including Yunhe East Street and other roads in the area, and features a total length of roughly 7.6 km (including 2.2 km of municipal roads).

The design team proposed to integrate the shaded path network, fitness trail network and sponge network to create an ecological framework for the pilot start-up area with an emphasis on integration. The team applied advanced concepts and technologies of "sponge cities", "smart gardens", "energy-efficient gardens" and "low-carbon gardens", integrated graded paths for environmental optimization, and focused on a safe and pleasant outdoor experience in creating 100% shaded walking paths and implementing the optimal positioning of the sub-center forest administrative office area so as to create an ecological garden and demonstration area for green innovation.

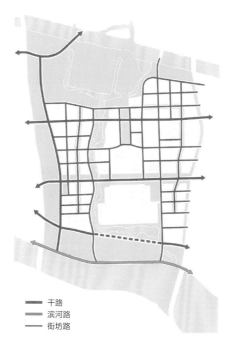

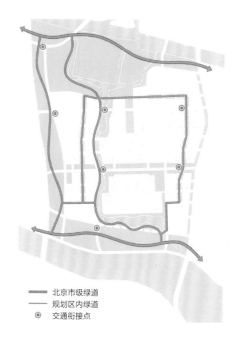

北京城市副中心先行启动区慢行系统优化

（1）营建端庄大气、清新明亮的林荫路景观

在道路绿化设计中，需要考虑与周边办公区绿地的关系，使用地性质有机结合，串联弹性场地空间，将道路绿化融入行政办公区整体规划中，打造端庄大气的林荫路景观。运河东大街贯穿行政办公区，与市委、市政府毗邻，连接六环公园游憩带、宋梁路生态风景带，是办公区内尤为重要的礼仪路及景观廊道。其他四条市政道路延续城市脉络，连接市委与委办局，实现行政办公区南北绿地与市政府东西绿地的自然过渡。

基于道路尺度、树种选择、四季季相、配置方式等综合考量，力求实现绿地最大化，以大乔木为主，辅以繁复多彩的地被花卉，营造舒适宜人的林荫交通环境，达到"清新明亮"的道路景观效果。采用具有北京特色的乡土树种，如市树国槐、市花月季、苍劲油松等，延续长安街等已建成道路的特色，结合植物季相变化，体现北京四季分明的景观特点，营造春花烂漫、夏荫浓密、秋色尽染、黛松映雪等富于变化的四季景色。

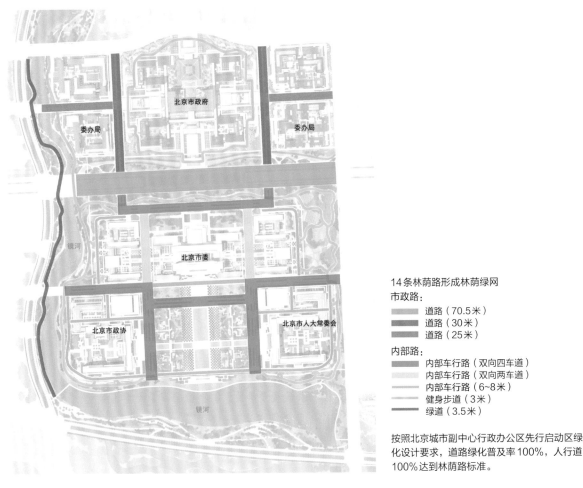

14条林荫路形成林荫绿网
市政路：
■ 道路（70.5米）
■ 道路（30米）
■ 道路（25米）

内部路：
■ 内部车行路（双向四车道）
■ 内部车行路（双向两车道）
■ 内部车行路（6~8米）
■ 健身步道（3米）
■ 绿道（3.5米）

按照北京城市副中心行政办公区先行启动区绿化设计要求，道路绿化普及率100%，人行道100%达到林荫路标准。

北京城市副中心先行启动区林荫路平面布局图

(1) Creating a Splendid and Invigorating Shaded Path Landscape

In terms of greening of the various walkways, it was necessary to consider the relationship with the surrounding office area's green spaces and organically integrate the site's existing nature, while incorporating a series of flexible spaces and the greening of walkways into the overall planning of the administrative office area to create a demure shaded path landscape.Yunhe East Street runs through the administrative office area, adjacent to the municipal party committee and municipal government, and linking the Sixth Ring Parks and Recreation Belt and Songliang Road Ecological Scenic Belt, which holds ceremonial significant and functions as a landscape corridor for the office area.The other four municipal roads extend the pulse of the city, connecting the municipal party committee and the commission, while achieving a natural transition between the green areas of the administrative offices from north to south, and the green areas of the municipal government from east to west.

Based on the dimensions of paths, selected tree species, seasonal phases, their configuration and other careful considerations, we strove to maximize the amount of visible green space, primarily through the use of mature trees, supplemented by a sophisticated and colorful flowering ground cover, to create a comfortable and pleasant shaded environment for foot traffic and achieve a "bright and refreshing" landscape effect.The use of native species which characterize Beijing, such as the *Sophora japonica Rosa chinensis* and *Pinus tabuliformis* perpetuate the features of iconic roads such as Chang'an Street. By integrating the seasonal changes reflected in these plants, the unique landscapes of Beijing's four seasons are more prominent, while embodying the vibrant spring flowers, summer shade, rich autumn colors, and winter pines which characterize each season.

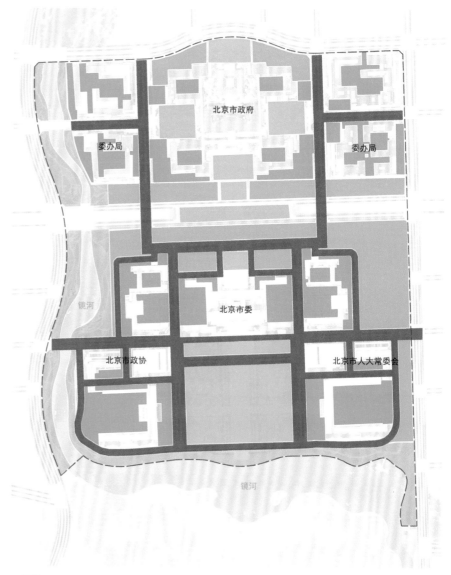

大树林
让建筑融于树林当中，提高环境舒适度，营造树林中的办公环境

外网绿环：先行启动区北侧起伏的绿地，复层种植的植物形成绿色屏障，环抱场地。
林荫绿网：统领全区的林荫路网，营造舒适的车行、步行空间。
生态轴：由景观和建筑共同构成的轴线，端庄大气、经纬交织。
绿庭：花园式的办公庭院为室外交流、健身、休憩提供尺度宜人、风景优美的空间。
绿带：镜河河道环绕在地块西南侧，自然式种植，让优美的自然山水风光融入先行启动区的办公环境当中。

北京城市副中心先行启动区种植结构平面布局图

达济街和通源街剖面图

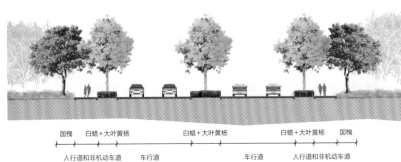

内部路 1 剖面图

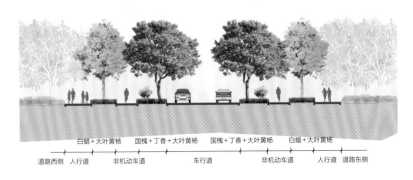

运河东大街辅路与绿地剖面图

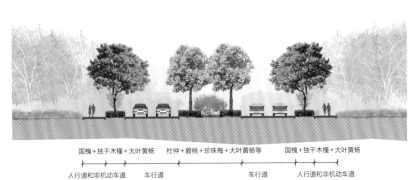

内部路 2 剖面图

清盛路和承安路剖面图

运河东大街林荫路网

承安路和清盛路的全林荫让炎热的夏天也变得更加舒适

多彩的林荫路网

通源街和达济街在各办公楼的出入口处设置了很多林下停车位

（2）一体化建设功能复合的多维林荫路网

不同于传统单一的绿化种植，本项目通过一体化设计营建先行启动区的多维林荫路网以及环网结合的立体绿道、健身步道系统。在划分道路等级的基础上，使林荫路设计满足车行、慢行、健身、海绵网以及照明、智慧交通等多种功能的需要。

区域内道路绿化需要统筹考虑车行、慢行要求，达到道路绿化普及率100%，人行道均为林荫路的标准。结合道路等级，对慢行道进行断面优化调整，创造安全、舒适的绿色出行体验。在机非分隔带预留较宽绿地，在人行道留双排树，保障车行交通顺畅的同时，为慢行系统营造舒适的林荫环境。

为满足办公室人群健康需求，在场地的"大树林"中植入林荫健身慢跑径，与公园绿道网一体化设计，将运动场健身、器械健身与无器械健身相结合，形成场地内的科学健身网络，引导科学健身，也有助于使用者缓解心理压力。

(2) Building a Cohesive and Multi-dimensional Shaded Network with Integrated Functions

Unlike traditional one-dimensional approaches to greening, a multi-dimensional shaded network, vertical greenway and fitness greenway system have been combined with the ring network through an integrated design.By classifying the network's various paths, the shaded path design can meet the needs of a range of functions such as vehicular traffic, leisure traffic, and fitness activities, while supporting a sponge network with lighting and intelligent traffic features.

The greening along the paths had to consider the needs of both vehicular traffic and leisure traffic to achieve 100% greening along the main paths and shaded pedestrian paths.Together with various path classifications, cross-sectional optimizations and adjustments were applied to the leisure paths, with an emphasis on safety, comfort and an environmentally-friendly experience.This involved setting aside wide green spaces in non-vehicular zones, with rows of trees lining the walking paths to ensure the smooth flow of vehicular traffic while providing a comfortable shaded environment for leisure traffic.

In order to meet the health and fitness needs of office workers, shaded jogging trails from the Great Woods were installed in the site and integrated with the park's greenway network. The design aimed to combine sports grounds, equipment-based fitness activities, and equipment-free fitness activities, so as to promote a scientific approach to health and fitness, guide users in improving their overall health, and help relieve psychological stress.

功能复合一体化的绿化设施带

健身步道穿行于林中和水畔之间，给人步移景异的景观体验

风景随着健身步道不断变换

绿道网串联湖景

健身步道沿着河道不断延伸

健身步道串联各种类型的城市阳台,使健身与休憩等活动融为一体

健身步道从杜梨林中穿过

河边林荫下的健身步道可随时欣赏河景

健身步道建立了城市阳台、建筑和景观之间的联系

057

健身步道在风景中起伏

从健身步道望向城市桥梁

洒满金叶的健身步道

结合道路绿网建设符合"海绵城市"理念的城市片区，使绿地实现海绵网全覆盖，三年一遇的雨水零外排，兼具雨水调蓄、补给地下水和乡土生境保护等多重功能。通过集雨型绿地、透水铺装、生态草沟等对雨水的收集、净化、滞留、渗透和积存，促进雨水资源的利用和水生态环境的保护。

在林荫路网从规划设计到施工完成的全过程中，设计团队与市政等部门统筹协调、综合考量、留出余量，在复杂的建设条件下闪转腾挪，全程跟进并及时协调与调整，保证了最终的建设效果。

By integrating a green network of paths into the concept of a "sponge city" within an urban context, the area's green spaces were able to achieve full coverage across the sponge network, while supporting functions such as zero outflow of rainwater over a three-year period, rainwater storage, groundwater replenishment, and the protection of the surrounding native habitat. The collection, purification, retention, infiltration, and accumulation of rainwater through the application of these rainwater-harvesting green areas, permeable paving and ecological ditches promote both the effective use of rainwater resources and the protection of the environment's aquatic ecosystem.

From the design and planning to the construction and completion of the shaded path network, the design team carefully coordinated their actions with the city and other relevant departments. In the face of challenging construction conditions, the team took every precaution into consideration by leaving a margin for error, following up, and coordinating and adjusting in a timely manner to ensure the success of the final construction.

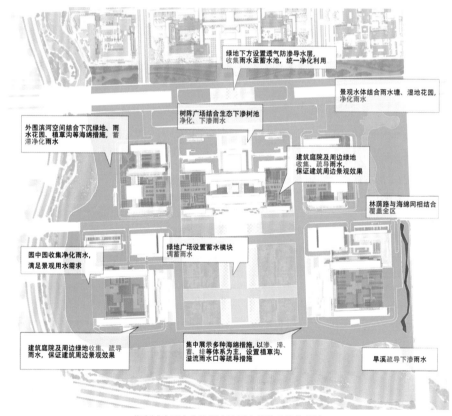

海绵网——蓄滞雨洪，雨水利用
建设符合"海绵城市"理念的城市片区，使景观兼具雨水调蓄、补给地下水和乡土生境保护等多重功能。通过对雨水的收集、净化、滞留、渗透和积存，促进雨水资源的利用和水生态环境的保护。

北京城市副中心先行启动区中的海绵措施设计

04 北京城市副中心行政办公区镜河河道景观专项设计
Landscape Design of the Beijing Municipal Administrative Center Jing River Area

项目地点： 中国 北京市通州区
项目规模： 38 公顷
设计时间： 2016—2018 年

Project location: Beijing, China
Project scale: 38 hectare
Design period: 2016-2018

镜河原名丰字沟，是连通运潮减河与大运河之间的排水渠道，从通州北京城市副中心行政办公区中央场地穿过，与北京老城血脉相连。

设计团队在北京市建筑设计研究院朱小地院长主持的北京城市副中心行政办公区总体城市设计指导下，将原丰字沟河道外迁，让出行政办公区的中央场地，改造为兼具休闲游赏、排水调蓄、生态节能等功能的城市风景河道——镜河，运用可游可赏的造景手法，融入智慧和科技创新等新技术，打造开放、共享、多功能的城市河道及滨水空间，激发市民低碳、乐活、健康的水畔生活。

The Jing River, formerly known as Fengzigou, is a drainage channel connecting the Yunchaojian River and Grand Canal. The river runs through the center of the Beijing Municipal Administrative Center in Tongzhou and shares a historical heritage with Beijing's old city.

Under the guidance of the Urban Planning of Beijing Municipal Administrative Center, chaired by President Zhu Xiaodi of the Beijing Architectural Design and Research Institute, the design team relocated the former Fengzigou River to make way for the central site of the administrative office area, transforming the area into an urban scenic river with a wide range of functions ranging from recreation and leisure, to drainage and retention, energy conservation, etc.. By implementing various design techniques to promote appreciable and leisurely attractions with the integration of new smart technologies and technological innovation, the team created an open, shared, and multi-functional urban river and waterfront space to inspire greater public engagement with a vibrant, healthy and low-carbon lifestyle.

丰字沟连通北运河与运潮减河（2006年卫星图）

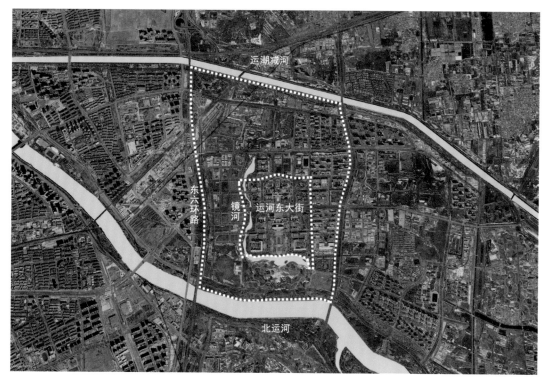

北京城市副中心水系（2022年卫星图）

水系是城市的经络，关系到城市的气韵。老北京城设计精密、功能复杂的水系在半个多世纪的城市建设、改造中遭到不同程度的破坏。北京城市副中心以创新的理念与设计，传承中国传统山水文化及营城理念，为北京城水系的再造提供参考。

通州与北京老城自元代起由运河、通惠河连通，百年以来同气连枝、一脉相承。北京老城内的水系集自然景观与人文景观于一体，体现了中国传统价值观与自然观，也孕育了北京城独特的地域人文气质。位于北京通州城市副中心的三条水系——运潮减河、镜河、北运河，承接了北京古城的水韵，延伸了古都文脉，作为润城之水滋养了新区，是副中心"北京气质"的载体。

The water system is the lifeblood of the city, and influences its character and temperament. The sophisticated design and complex functions of the old Beijing city water system had been damaged to varying degrees during more than half a century of urban construction and transformation. Introducing a fresh concept and innovative design, the Beijing Sub-center inherits the traditional culture of Chinese landscapes and Yingcheng (ancient city planning), while also providing a model for the re-creation of Beijing's water system.

Tongzhou and the old city of Beijing have been connected by the Grand Canal and Tonghui River since the Yuan Dynasty, and have converged together for centuries. The water system within the old city of Beijing combines natural and cultural landscapes, which reflect traditional Chinese values and concepts of nature, as well as nurturing the unique regional culture and character of the city. The three water systems in the Beijing sub-center of Tongzhou—the Yunchaojian River, Fengzigou River and North Canal—inherit the charm of the ancient city of Beijing and expand its cultural lineage, infiltrating the sub-center in the same vein as the storied waters of the city, and conveying the traditional spirit of Beijing.

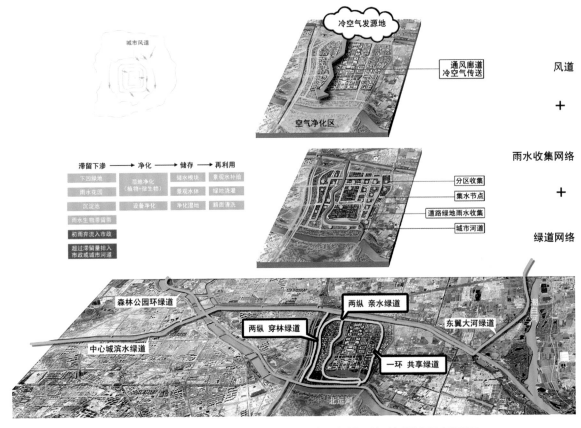

一体化设计，"多道合一"，实现镜河河道复合多元的功能，达到土地资源的高效利用

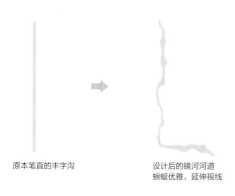

原本笔直的丰字沟　　设计后的镜河河道
　　　　　　　　　　蜿蜒优雅，延伸视线

1. 区域水系梳理结合城市空间和景观视线，塑造蜿蜒优雅的水系形态，引导视线延伸，尺度适宜，创造放松舒适的滨水景观。

2. 依据人体视觉原理，选择最佳观赏角度，设计道路及休闲平台，丰富视觉体验。结合岸线与种植设计，使河道景观富有节奏变化。

3. 河道最宽处百余米，最窄处约20米。

依据视觉设计构建镜河水体形态　　　镜河各区段长宽尺度　　　镜河河道关键节点与视线分析

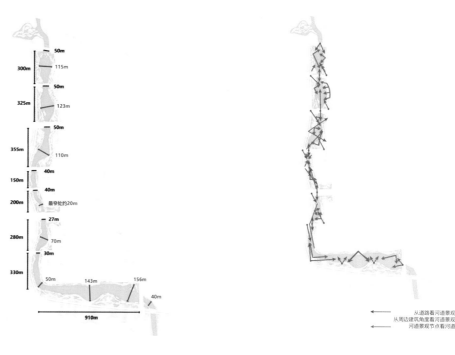

　　景观的营造从来都是系统工程。镜河河道的规划设计以跨领域、多专业的统筹协作为预设条件。这里的关键词是"一体化"。水系的设计必须注重总体效果，强调各局部、细节间的呼应。在相互映衬、呼应又富于整体感的开放空间中，使绿色环保、低碳节能、舒适宜居得到最大限度的实现。

　　镜河河道的设计规划，以有利于共享为设计目标，充分考虑周边居民的需求，景观设计处处体现亲和、宜人。经由水岸线、滨水绿化观景网、健康休闲绿道等，打造开阔、兼容、多功能的公共空间，使水资源及周边环境得以高效利用。

　　水系与滨水景观的营造，不但是北京城市副中心整体景观的一部分，而且将成为其灵魂，并辐射影响周边地区。设计团队的愿景是"在不同专业的协同努力下，达到综合效益最大化，使节约型的生态园林景观与建筑融合一体，自然与人工无缝衔接，将所在区域建成历史文化与现代文明交汇的新型城区的样本"。

　　The creation of a landscape is much more than just a systematic undertaking. The planning and design of the Jing River was premised on cross-disciplinary and multi-disciplinary collaboration. The key word here is "integration". The design of the water system focused on the overall effect of the feature, emphasizing the relationship between respective areas and detailed elements. By creating an open space which reflects and echoes these rich features as a whole, the design team were able to maximize the environmental protection, energy conservation, and overall comfort of the area to the greatest extent possible.

　　The design and planning of the Jing River was pursued in a way that was conducive to mutual sharing, taking into full consideration the needs of the surrounding residents and the landscape to create a friendly and pleasant environment. By utilizing the waterfront, scenic green network, and leisure greenway, the team created an open, compatible and multi-functional public space that makes efficient use of the available water resources and surrounding environment.

　　The water system and waterfront landscape will function as not only a part of the overall landscape of the Beijing Municipal Administrative Center, but the very soul of the site which exerts a direct influence on surrounding areas. The design team's vision was to maximize potential benefits through the concerted efforts of various disciplines, so that ecological landscapes and architecture, and to a greater extent nature and humanity, may be seamlessly integrated, so as to create a new model urban area where history and culture meet modern civilization.

(1)"多道合一"实现资源高效利用

镜河河道景观设计从统筹协调的角度出发，与相关的水利、交通、能源、生态环保等多专业合作，运用一体化设计的手法，将地源热泵、水道、绿道、风道、节地设计"多道合一"，统筹考虑水质提升、资源可再生及生物多样性等需要，将生态水岸、集雨型绿地、低能耗绿地一体化营建，打造复合多元的湿地生态系统，塑造北方城市典型的风景河道景观，促进低碳发展，实现土地资源的高效利用。

景观设计顺应北京盛行风向，塑造包括镜河河道和滨河绿地约6平方公里的通风廊道，促进空气微循环，缓解热岛效应，结合降温增湿的辅助设施，提升环境舒适度；预先考虑镜河河道下埋地源热泵管线的需要，此管线提供周边建筑冬季供热、夏季供冷的生态能源，减少碳排放；构建集雨型绿地，形成"城市海绵"，保证三年一遇无外排；收集滞留雨水释给绿地，节约浇灌，提高土壤及空气湿度；构建低能耗绿地，通过景观构筑物和活动场地，结合太阳能光伏发电装置，应用可再生能源实现景观零能耗。

对原有的笔直岸线进行生态水岸改造，形成4公里形态蜿蜒曲折的自然水岸，配植水生、湿生植物，构建水岸生态系统，结合其他措施净化滞留雨水，提升河道水质。通过稳定的植物群落结构吸引鸟类、鱼类、昆虫栖息，有效提升河道的生物多样性，发挥生态效益，起到消减噪声、降尘固碳、降低PM2.5等作用。

(1) "Multi-channel Integration" to Efficiently Use Resources

The landscape design of the Jing River is predicated on cross-disciplinary and multi-disciplinary coordination, and focuses on the overall effect of the feature, emphasizing the relationship between respective areas and detailed elements. By creating a ribbon-shaped space which reflects and echoes these rich features as a whole, the design team was able to realize multiple functions through multi-feature integration to promote low-carbon development and the efficient use of land resources.

With regards to the project's coordination, we cooperated with relevant partners in relation to water conservation, transportation, energy, ecological and environmental protection, etc., and applied an comprehensive approach to integrating a geothermal heat pump, waterway, greenway, windway and environmental design. Taking into account the need for improved water quality, renewable resources and greater biodiversity, the team implemented an ecological waterfront, featuring energy-conserving green spaces that promote rainwater collection, thus creating a composite and diversified wetland ecosystem and model for scenic river landscapes in northern Chinese cities.

Based on the prevailing wind conditions in Beijing, a ventilation corridor of roughly 6 square kilometers including the Jing River and riverfront has been shaped to promote air microcirculation and alleviate the impact of urban heat islands, which has been combined with auxiliary facilities related to cooling and moisture to enhance the overall comfort of the environment. The landscape design takes into account the need for a geothermal heat pump system under the Jing River to ecologically address the energy needs of the surrounding buildings with regards to heating in winter and cooling in summer, while reducing carbon emissions. In addition, the team built rain-collecting green areas that form "urban sponges" to ensure no external drainage over a three-year period, and collected and reallocated stagnant rainwater to green areas to both reduce the need for watering and improve soil and air humidity. Energy-saving green spaces were constructed by implementing landscape structures and activity sites combined with solar photovoltaic power generators, through which zero energy consumption was achieved across the entire landscape.

The original straight shoreline underwent an ecological transformation involving the creation of 4 kilometers of meandering natural waterfront, with aquatic plants and wet plants, so as to create a waterfront ecosystem which, in combination with other measures, facilitated the purification of stagnant rainwater and improved the water quality of the river. The sustainable plant habitat attracts birds, fish and insects, and effectively enhances the biodiversity of the river by applying ecological benefits and playing a role in noise abatement, dust and carbon reduction, and reducing PM2.5 levels.

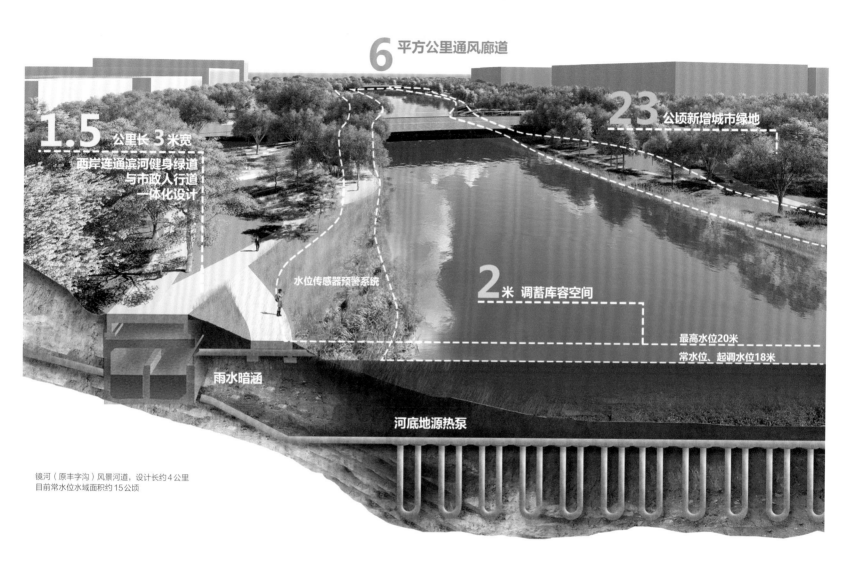

镜河设计原则1——功能复合的风景河道：地源热泵、水道、绿道、风道，"多道合一"

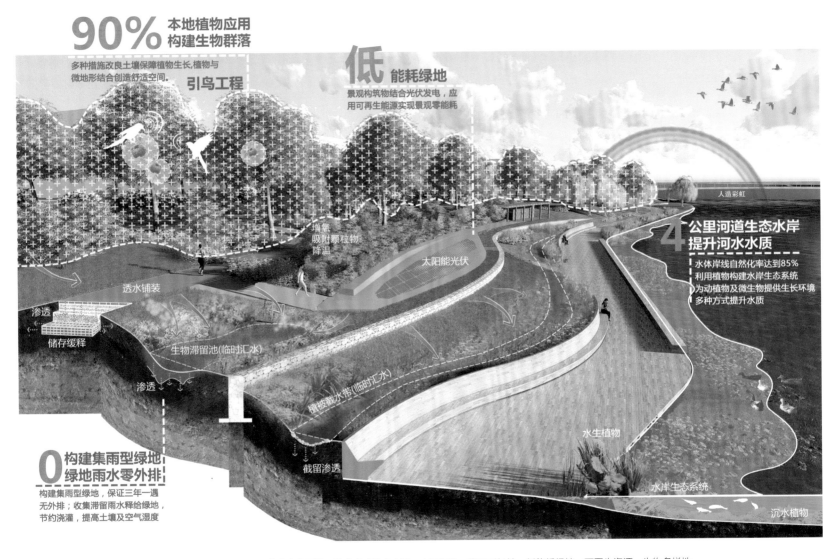

镜河设计原则2——一体化营建生态系统：综合考虑生态水岸、水质提升、集雨型绿地、低能耗绿地、可再生资源、生物多样性

镜河设计原则3——创新宜居的城市生活：通过滨河景观、林荫环境、智能设施营造舒适小气候，提升宜居指数

镜河设计原则4——多层次的绿色开放空间：休闲空间系统、科学健身系统、科普教育系统、公共艺术系统

运河东大街北侧是河道中最窄处，集合了地源热泵、水道、绿道、风道、节地设计等"多道合一"，体现了包容、开放的理念

城市的开放空间与景观河道融为一体

运河东大街南侧镜河河道景观，体现出安静而生态的办公环境氛围

城市河景成为北京城市副中心的标志景观

办公区被镜河水系风景环绕

初春,百米宽的镜河水岸为生物提供了更多的栖息之地

镜河景观二期工程营建的百米宽镜河水面形成的湖景景观

河道的最窄处

百米宽的镜河水岸的初春景观

早春的镜河风景

利用溢流口等地下设施顶部形成近百米的亲水平台,激活水岸生活

早春镜河景观(一)

早春镜河景观(二)

在百米亲水平台可边钓鱼边欣赏对岸台地景观

夏季,百米宽的湖面成为水鸟栖息之所

初春风景下的办公建筑群

初春的多彩景观倒映在河道中,形成了一幅色彩斑斓的画卷

建筑、河道与秋景植物形成了和谐统一的风景

初秋河道傍晚景观

夕阳映衬下的办公建筑群

从观景平台向对岸望去

在观景平台上可眺望到各式各样的桥

闲暇之余可以在岸边遛弯与闲谈

丰富的植被形成自然的河道景观

将河道风景引入城市景观

站在城市阳台远眺建筑群

夕阳下的城市阳台

各式各样的亲水平台

夕阳下的亲水平台

亲水平台作为接触自然与健身休憩的最基本单元嵌入风景河道中

站在城市亲水平台上看委办局办公楼

坐在亲水平台上观夕阳

生态驳岸（一）

生态驳岸（二）

生态驳岸（三）

（2）场地空间传承文化基因

镜河河道景观设计借鉴中国古典园林中的理水思想，结合现代城市开放空间与风景河道的设计手法。作为承接北京古城的水韵和文脉、传承"北京气质"的重要载体，其景观设计既借鉴皇家园林的恢宏阔大，又融入胡同文化的平易亲和，于潜移默化中将景观精细化设计与文化性、实用性、节材节地相结合，为避免千城一面、景观同质化提供全新的路径。

以润物细无声的方式将文化融于场地中，结合使用功能与建造需要，突出表达北京老城对副中心的文化渗透。通过材料选择、建造方式、施工技法等具体细节，展现北京数百

(2) Inheriting the Cultural Character of the Site

The landscape design of the Jing River waterway draws on the concept of water management in classical Chinese gardens, in combination with the intelligent integration of modern urban open spaces and scenic rivers. As an important conveyor of the charm and cultural lineage of the ancient city of Beijing and its historical character, the landscape design draws on not only the grandeur of the royal gardens, but also the friendly and down to earth culture of the hutongs. By subtly infusing a refined landscape design with cultural and utilitarian elements, and implementing resource and land conservation, the design team has provided another approach that differs from homogeneous cities and landscapes.

By discreetly integrating cultural elements, and combining the use of functions and construction needs, the site highlights the integration of Beijing's old city with the sub-

借鉴中国古典园林的思想，随坡就势、叠石筑岸的秋景

北京城市副中心京味十足的冬日河景

年间城市发展的线索与历史文化的脉络，储存北京民众共同记忆的情感线，让场地在满足景观功能的同时成为城市文化基因的载体，蕴含独特的感染力。

在镜河水畔、市委办公区南广场两侧绿地中，保留了场地拆迁前村中的椿树、榆树等老树，延续了场地的记忆与脉络。沿河的植物设计结合中国北方传统园林中的经典植物场景，移步异景，重现了朱自清、郁达夫等文人笔下北京那些令人难忘的场景，同时巧妙应用颐和园、北海等传统园林里的经典植物场景，让使用者感到熟悉亲切，传递北京气质。

center, and preserves the connections to Beijing's urban development, history and culture over the centuries through specific details such as material selection, and construction methods and techniques, thus preserving the emotional ties that bind together the common memories of Beijing residents and empowering the site as a unique and compelling bearer of cultural legacy, while satisfying the necessary functions of the landscape itself.

The green spaces around the Jing River waterfront and South Plaza feature old trees such as *Toona sinensis* and *Ulmus pumila* from the village, which were preserved before the site was demolished, so as to perpetuate the memory and legacy of the site. The landscaped plants along the river combine classic plantscapes from traditional gardens in northern China, so as to recreate those unforgettable, moving scenes of Beijing as described by literary figures such as Zhu Ziqing and Yu Dafu, while delicately blending the historical plantscapes of traditional gardens such as the Summer Palace and Beihai, to bestow the office area with a familiar and friendly tone that conveys the character of Beijing.

水岸营建体现了中国传统园林"叠山理水"的造园手法

中国传统叠石造园与生态驳岸相互融合

资料链接4

《HOME.安住》镜河健身慢跑径封边艺术装置作品

《HOME.安住》是一件长约1.6公里、宽30厘米的大型当代艺术装置作品，由著名艺术家黄海涛领衔创作，无界景观设计师与北方石匠共同参与完成。作品在满足镜河河畔健身慢跑径封边石（道牙）功能的同时，承载展现北京数百年间城市发展线索的文化功能。

作品由镌刻着千余条北京胡同名字的石料以榫卯的方式契合衔接，形成一条储存北京民众共同记忆的情感线，具备激活游人的心理记忆、为运动休闲场所（镜河绿道）创造出独特感染力的功能。

作品材料是具有北方城市代表性的建材片段，如城墙砖、旧砖、老汉白玉、青石以及花岗岩等。通过传统榫卯的方式相互契合成组，使公众的个人记忆在此汇聚成为共同的记忆之河，产生心理共鸣，增强行经的市民对于自己城市的认同与归属感，让场地在满足景观功能的同时也能够成为城市文化基因的载体。

作品在记载胡同名称的同时，也记录了部分胡同名称在数百年间的演化，储存了丰富的历史信息。设计团队在健身慢跑径封边上共镌刻了千余条胡同的名称，以不同的地区进行区分，并将胡同的曾用名以缩小字体的形式出现在了胡同现用名的周围。

Link 4

HOME Jing River Fitness and Jogging Trail Edge Installation

HOME is a large-scale contemporary art installation that is roughly 1.6 km long and 30 cm wide. It was led by the famous artist Huang Haitao, with the participation of View Unlimited Landscape Design and Northern Stonemasons. The work fulfills an important function, namely sealing the stones (pavement) of the fitness and jogging trail along the Jing River, while also conveying the culture of Beijing's centuries-old urban development.

The work is composed of stones engraved with the names of more than a thousand Beijing hutongs, which are joined together in a mortise and tenon design to form an emotional lineage that preserves the common memories of Beijing residents, while inspiring visitors and creating a unique and compelling environment for sports and leisure (the Jing River Greenway).

The materials used in the work are typical of northern cities, including materials such as city wall bricks, aged bricks, Chinese white jade, limestone, and granite, etc. By applying a traditional mortise and tenon method, each piece fits into each other, creating a confluence of personal memories that converge together, and a resonance that enhances each citizens' sense of identity and belonging to the city, while allowing the site to both convey the city's cultural legacy and satisfy the necessary functions of the landscape itself.

In addition to recording the names of hutongs, the work also records the evolution of certain hutong names over the centuries, thus storing a wealth of historical information. The design team engraved a total of more than a thousand hutong names that paved the fitness and jogging trail, which were distinguished by their respective areas, and arranged the former names in a reduced font around the current names of the hutongs.

健身慢跑径封边艺术装置阅读指南

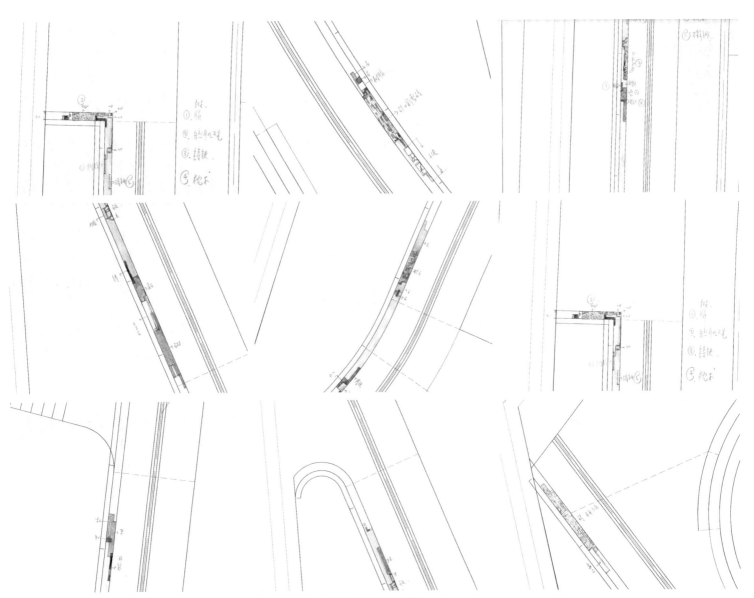

艺术家黄海涛老师手稿

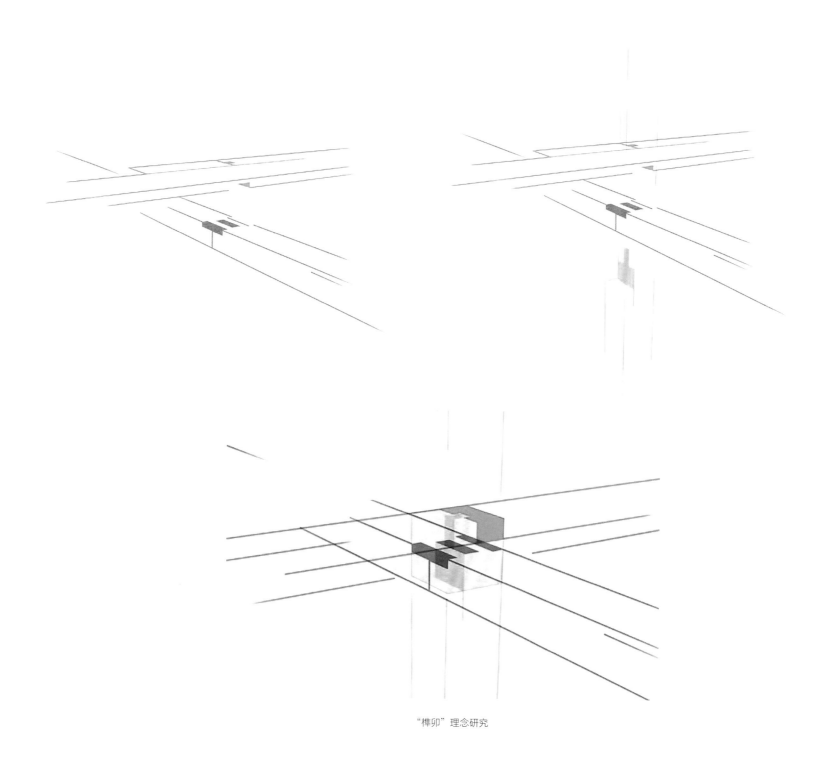

"榫卯"理念研究

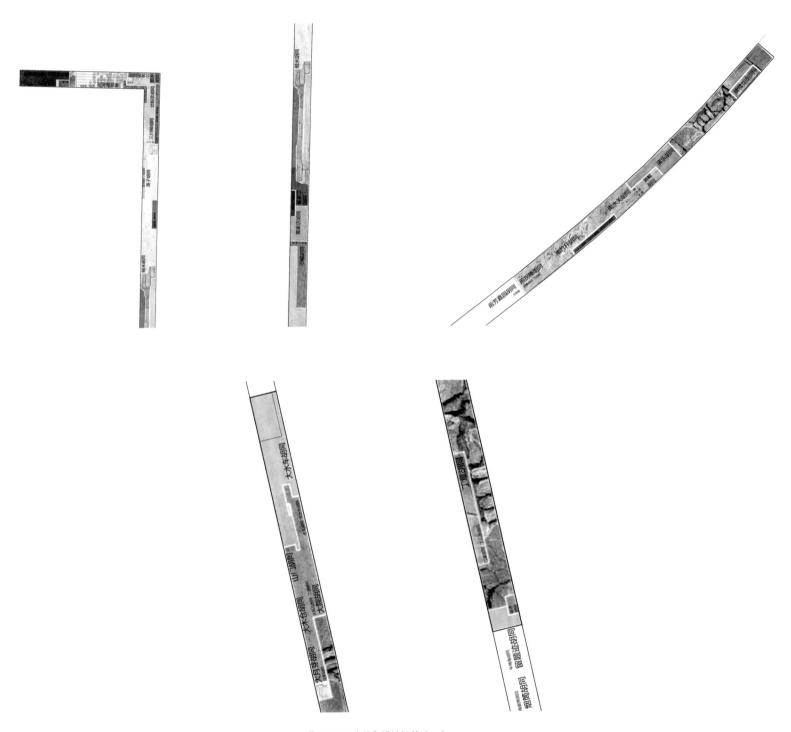

《HOME．安住》设计细节（一）

《HOME.安住》设计细节（二）

该作品是无界景观工作室于2016年入选第15届威尼斯建筑双年展中国国家馆的大型综合装置作品《安住·平民花园》的延伸。《安住·平民花园》展现了无界景观工作室数年间于北京老城区杨梅竹斜街展开的环境更新实践，记录了无界景观工作室从公共设施建设到引导居民通过共建胡同花草堂，形成自发共享社区的经验❶。双年展之后，设计团队将部分展品进行了重铸，制成双年展砌块砖融入镜河健身慢跑径的艺术装置中，期望北京胡同活力化这一缓慢发展、持续发酵的课题能够得以延续。

The large-scale installation HOME-COMMUNAL GARDEN by View Unlimited Landscape Architecture Studio was selected for the 15th Venice Architecture Biennale in 2016 and was featured in the Chinese Pavilion. HOME-COMMUNAL GARDEN showcases the environmental regeneration practices of View Unlimited Landscape Architecture Studio in Yangmeizhu Xiejie in Beijing's old city. It documents the experience of View Unlimited from the construction of public facilities to guiding residents in forming a spontaneous shared community through the co-creation of Flora Cottage❷. After the Biennale, the design team reconstituted some of the exhibits from the Biennale into bricks to be incorporated into the Jing River Fitness and Jogging Trail art installation, with the express purpose of furthering the slowly evolving and continuously percolating subject of the revitalization of Beijing's hutongs.

❶ 参考资料："中国公共艺术丛书"第一册《安住·杨梅竹斜街改造纪实与背后的思考》童岩，黄海涛，谢晓英著。
❷ Reference: HOME-Home Documentaries and Thoughts Behind the Reconstruction Project of Yangmeizhu Xiejie by Tong Yan, Huang Haitao, and Xie Xiaoying.

约1.6公里长的健身慢跑径封边镌刻了干余条北京胡同名字,形成一条能够储存北京民众共同记忆的情感线(部分由耿毅军先生摄影)

（3）风景河道融入日常生活

镜河河道景观设计将滨河的风景林荫路网与市政人行道一体化设计，营造绿树成荫的河道景观。充分考虑办公人群与周边居民的需求，以利于共享为原则，为人们提供可游可赏、亲近自然、健身休闲的场所，让河道融入日常生活。丰富的水体形态变化，结合周边建筑环境分区段设计的富有创意的独特景观亮点，共同形成开合有致、灵动大气、整体统一又富于变化的滨水空间，激活水岸生活，提升宜居指数。

(3) Integrating Scenic Rivers into Everyday Life

The Jing River landscape design adheres to the concept of green development and features a tree-lined scenic river. The riverfront's scenic shaded path network and municipal sidewalks feature an integrated design which takes into full consideration the needs of the office population and surrounding residents, while ensuring that the design is conducive to shared access by providing places to participate in fitness and leisure activities in close proximity to nature, thus integrating the river into everyday life. The rich changes in the waterscape and creative and unique landscape attractions, which have been designed in combination with the zoning requirements of the surrounding environment, together form a waterfront space with a coherent and dynamic atmosphere that is unified and yet diverse, and preserves the lifestyle of the waterfront while enhancing the overall comfort of the area.

午休时上班族在镜河边运动

广场成为老年人晒太阳观风景的地方

注重滨水可达性，通过河道自然高差和局部微地形改造，在滨河绿地内植入形式多样、尺度适宜的休闲空间，营造景观节点即"城市阳台"，具体包括亲水平台、栈道、台阶广场、带状林荫休闲空间等，激活水岸生活，为风景河道叠加多样化的使用功能。注重水岸两侧建筑与河道水景的互动关系，保证林荫路的滨水景观视野，合理布局开放及半开放空间，形成室外交流、亲水休闲、科学健身、科普教育、公共艺术等景观化、多样化的开放空间系统。

With a focus on waterfront accessibility, and by harnessing the natural height difference between the river and changes in local micro-topography, a diverse riverfront green space has been introduced, which features a scaled recreation and leisure space, so as to create a landscape node akin to an "urban terrace". It includes waterfront decks, trestles, stepped plazas and shaded recreational spaces, with the goal of revitalizing the lifestyle of the waterfront and extending a range of more diverse features that compliment the scenic river. The design team paid special attention to the interactive relationship between the buildings on both sides of the waterfront and the river waterscape, to ensure that the shaded paths offered stunning waterfront landscape views, and integrated a rational layout of open and semi-open spaces to form a landscaped and diversified open space system for outdoor interactions, water-based leisure and recreation, health and fitness, science education, public art, etc.

秋景与秋叶也成为人们美好生活的一部分

镜河河道种植设计遵循因地制宜、适地适树、提高生物多样性的原则，注重植物的空间搭配和季相变化，既有高大乔灌木层彩叶植物的点缀，又有林下冷季型草坪和野花地被的花草结合，多方位发挥植物造景的优势。河道的植物采用功能复合型种植设计，在乡土品种的基础上，加强槭树、杂交马褂木、蓝杉等新优园林植物品种的应用。乔灌草合理配置，出现乔、灌体量与整体效果不协调时，及时调整，完善乔灌木的空间感，去繁化简，削弱留强，优化植物组团，使

The plantscapes along the Jing River follow the principles of localization, adaptation, and improvements to biodiversity, with a focus on spatial collocation and seasonal changes. Tall trees and shrubs are accented with foliage, while a combination of flowers and grasses form the basis of cool-season lawns and wildflower ground cover in the forest showcase the advantages of plantscapes in a variety of ways. The plants adorning the riverbank have been designed to offer integrated functions, which have been allocated on the basis of native species, and reinforced with *Acer ginnala*, *Liriodendron chinense*, *Picea pungens* and other new garden plant species that are advantageous to the landscape. A sensible layout has been applied to the trees, shrubs and grasses that

市民在散步之余顺便停留在河边拍摄美景

午休的工作人员在河边平台放松身心

市民已逐渐把片区内的绿地作为周末的休闲之所

得植物组合尺度合适，层次自然过渡，视觉舒适，满足办公区内使用者及周边居民、游客的观赏需求。

沿镜河及办公区内绿地布置8公里长的蜿蜒流畅的健身游径，穿行于林中与水畔之间，创造步移景异的景观体验。设置无器械即时健身系统，引导办公人群及周边居民在优美的户外景致中随时随地使用，将健身活动融于风景之中，帮助人们舒缓压力、增进交流、提高工作效率、改善亚健康状态。

adorn the landscape, with timely adjustments to the volume of trees and shrubs as needed to maintain a coordinated overall effect which optimizes the perception of space, and imbues a sense of minimalism. Regular prunings and systematized plant groupings allow each combination to remain at an appropriate scale, with natural overlaps and a sense of fluidity that is pleasing the eye, so as to meet the viewing needs of office workers, visitors and surrounding residents.

An 8-kilometer-long fitness trail has been designed to incorporate the Jing River and green spaces within the office area. As the trail passes between the forest and waterfront, it offers users a different and unique landscape experience at every step. In combination with accessible freemotion fitness systems, the environment inspires the office population and surrounding residents to participate anytime and anywhere in the beautiful outdoor scenery. Furthermore, integrating fitness activities into the landscape provides people with a vehicle for stress relief, improved social interactions, an increase in work efficiency and reduced health risks.

市民特意来到场地内开启直播模式

美丽的上班路

周边的市民盛装打扮来此拍照

水边的平台成为市民平日聊天的场所

镜河河谷成为遛娃的好去处

林下的闲谈之处

城市边的广场是市民日常聚会聊天的理想之所

在镜河河畔花丛中钓鱼成为了一种生活方式

在风景中健身

在风景中休憩的市民

下车驻足欣赏城市秋景的市民

市民驻足拍摄美丽秋景

资料链接5

"软组织"北京城市副中心行政办公区镜河即时健身系统与公共艺术策划

Link 5

"Soft Tissue" Jing River Accessible Fitness System and Public Art Planning for Beijing's Sub-center Administrative Offices

"软组织·即时健身系统"的设计源于无界景观工作室在北京大栅栏杨梅竹斜街进行的环境更新与公共空间营造这一持续性的实验性项目。2017年北京国际设计周期间,无界景观工作室在北京杨梅竹斜街上策划"软组织·胡同中的即时健身系统"活动,利用墙面、台阶、花池坐凳等狭窄细碎的胡同街巷空间引导居民随时随地健身,并用及时贴(剪纸)、木板蚀刻等方式引导科学的无器械健身方法。这是一套软性的、安全的、温和的、能够随时随地使用的健身系统,使居民能够利用身边的构筑物进行最基础的身体拓展训练,让科学健身的新方法进入老城最深处的细碎空间,到达城市的"末梢神经",辐射缺乏健身意识的人群,引导绿色、健康的生活。

在这之后,无界景观工作室把"软组织·即时健身系统"的理念从北京老城中心——大栅栏的杨梅竹斜街,移植到北京通州的新中心——行政办公区的镜河沿岸。

The design of the "Soft Tissue-Accessible Fitness System" originates from the ongoing experiments in environmental regeneration and public space creation by View Unlimited Landscape Architecture Studio in Yangmeizhu Xiejie, Dashilan, Beijing. During the 2017 Beijing Design Week, View Unlimited planned the "Soft Tissue-Accessible Hutong Fitness System" in Beijing's Yangmeizhu Xiejie. The system utilized the narrow and fragmented space of the hutongs, including walls, steps and flower beds, in order to motivate residents to work out anytime and anywhere, and used relevant stickers (paper cutouts) and wood etchings to guide residents in scientific fitness activities without equipment. By offering a mild, safe and gentle fitness system that can be used anytime and anywhere, residents were empowered to use everyday structures around them for their most basic fitness training needs, allowing new scientific methods to take root in the deepest and most fragmented spaces in the old city and penetrate into the "peripheral nerves" of the city, so as to raise people's awareness of physical literacy and promote green and healthy lifestyles.

After this experience, View Unlimited transplanted their conceptual "Soft Tissue-Hutong Fitness System" from Yangmeizhu Xiejie in Dashilan, the heart of Beijing's old city, to Tonzhou's new sub-center, along the Jing River municipal administrative area.

依托景观台阶随时随地健身

台阶与看台结合,提供弹性休息空间

在北京杨梅竹斜街一名法国建筑师正在跟着无器械健身标牌做运动

充分利用镜河河道两岸带状绿地的空间形式，以及市政道路与河岸存在竖向高差的优势，塑造美丽的滨河景观，合理配植植物并结合智能设施，形成低于市政道路的，降温、降噪、除尘的小气候空间。因地制宜地利用场地内的绿道、坡道、台阶、栏杆、灯杆、座椅等设施，布置河道沿岸"软性、安全、温和"的无器械即时健身系统，配以金属牌标识，介绍使用方法。结合连贯的健身步道、休憩平台、避雨廊架等，引导办公人群及周边居民在优美的户外景致中随时随地健身使用。

未来计划结合智能化的网络科技系统，将科学健身、自然科普、文化艺术等信息与手机终端相关联，实现"边走边玩、边玩边学"，引导科学合理的健康理念，增进人与人之间的交流互动，培养健康的生活方式。

无器械健身系统不仅让住在杨梅竹斜街胡同里的居民能跟着健身标识运动，也能吸引来到这里的游客运动打卡

ANTI 抗体 BODY

各类无器械健身标识图纸之一

拉伸练习 STRETCHING EXERCISES

大腿 THIGH

股外侧肌 gǔ wài cè jī VASTUS LATERALIS

时间：5~6秒
动作要领：
支撑腿脚尖向前，膝盖伸直，用另一侧手抓住拉伸腿的脚踝处，向上缓慢拉伸，上体把腰绷直，挺胸。
辅助工具：
立柱，栏杆（提供不同高度支撑，因人而异），高度要求略比腰高，稳定性好。

《抗体·风景融入日常生活－即时健身系统》
作品入选第四届中国设计大展暨公共艺术专题展

By making full use of the spatial form of the green spaces on both sides of the Jing River, and the height difference between the municipal roads and river bank, a stunning riverfront landscape was crafted that combines optimal planting techniques with intelligent facilities to form a microclimate that in comparison to the municipal road is cooler, quieter, and free of pollution. The site uses greenways, ramps, steps, railings, light poles, seats and other facilities according to local conditions, and features accessible freemotion fitness systems along the river, with metal signs that provide instructions for use. A combination of interconnected fitness trails, resting platforms and rain shelters inspires office workers and residents in the surrounding area to engage in fitness activities anytime and anywhere amidst stunning outdoor scenery.

Future plans to combine intelligent network systems with science-based health and fitness, natural science, culture and the arts, and other information associated with cell phone terminals, aim to achieve a "Learn as you go, play as you learn" approach that promotes scientific and rational health concepts, increased social interactions, and the cultivation of healthy lifestyles.

各类无器械健身标识在办公区庭院场景中的应用——上肢及下肢部分

各类无器械健身标识在镜河河道与办公区内场景中的应用——全身肩颈部分

各类无器械健身标识在镜河河道场景中的应用——全身部分

2020年，无界景观工作室再次把"软组织·即时健身系统"理念从中国北京延伸到了埃塞俄比亚首都亚的斯亚贝巴。在亚的斯亚贝巴城市中心——谢格尔公园的景观设计项目中，因地制宜地利用公园内的游径、坡道、台阶、栏杆、灯杆、座椅等设施，布置遍布全园的无器械即时健身系统，为游客提供融于宜人风景中的休闲健身空间。配以多种语言介绍使用方法，通俗易懂，简单易学。该设计以舒适宜人的健身空间建立人与人之间交往的媒介，增进人与人之间的交流互动，提升社会的友好度及幸福感；同时，低成本、无器械的即时健身系统为埃塞俄比亚乃至全非洲的城市贫民聚集区以及村庄提供了可行的范例，为培养健康生活方式、促进民众身心健康、提高人口综合素质提供了可以大范围推广的模式。归根结底，民众的身心健康是社会繁荣发展的基本保障。

从老城胡同到新城河道，再到城市中央公园，无界景观工作室将"软组织·即时健身系统"由国内传播到了国外，由中国的首都传播到了非洲的中心（亚的斯亚贝巴是非盟所在地）。从景观专业角度来说，该系统体现了景观设计的精细化、人性化理念，以及在城市更新发展中的积极作用。从社会人文角度来说，则期望借由融于美丽宜人环境的、随处可见的即时健身系统，舒缓人的压力，疗愈身心，增加人与人之间的相互交流，搭建人与人交往的媒介，从而引导健康的生活方式，为促进城市与社会的稳定发展提供一种创新的可广泛推广的模式。

In 2020, View Unlimited Landscape Architecture Studio once again expanded its "Soft Tissue-Accessible Fitness System" concept from Beijing, China to Addis Ababa, Ethiopia. The landscape project for Sheger Park, located in the urban center of Addis Ababa, features trails, ramps, steps, railings, light poles, seats and other facilities that have been appropriately integrated to provide visitors with a leisure and fitness space that blends into the beautiful scenery by arranging accessible freemotion fitness systems throughout the park. The systems are easy to understand and learn, and support instructions in multiple languages. The design provides a comfortable and enjoyable fitness space and establishes a medium for increased social interactions and an enhanced sense of happiness and community. At the same time, low-cost, accessible instant fitness systems provide a feasible model for impoverished urban areas and villages in Ethiopia and Africa as a whole, and offer a model for cultivating healthy lifestyles, promoting physical and mental health, and improving the overall character of the population, which can be replicated on a large scale. Ultimately, the healthy and harmonious development of each individual's body and mind is the most basic guarantee for a prosperous society.

From the hutongs of the old city to the Jing River in Beijing's sub-center, and to Sheger Park in Adis Ababa, View Unlimited Landscape Architecture Studio is spreading its concept of a "soft tissue-accessible fitness system" from home to abroad, and from the capital of China to the very heart of Africa and the seat of the African Union. From the perspective of landscape professionals, the system embodies the concept of refinement and humanization in landscape design and its positive role in urban renewal and development. At the same time, from a socio-humanistic perspective, we are excited to see accessible fitness systems that are integrated into beautiful and inviting environments, and which can relieve stress, heal bodies and minds, increase mutual exchanges between people, and create a medium for human interaction, thus encouraging healthy lifestyles and providing an innovative and widely replicable model for promoting the stable development of cities and societies.

培养人们热爱运动的习惯

谢格尔公园内随处可见慢跑的人群

以在北京杨梅竹斜街、城市副中心镜河河岸以及埃塞俄比亚首都亚的斯亚贝巴谢格尔公园的即时健身系统为主题的《抗体·风景融入日常生活–即时健身系统》作品入选了第四届中国设计大展暨公共艺术专题展，无界景观设计团队通过研究分析疫情期间的社区连接，从连接美学和关系美学的维度上思考后疫情时期的公共艺术发展。

随着新冠疫情的到来，人们的社交活动大幅减少，健身空间持续受到挤压，因此即时健身系统为促进普通民众身心健康、培养健康生活方式、提高人口综合素质提供了可以大范围推广的模式，为城市人口聚集区及村庄提供了可行的范例，为社会的和谐发展提供了保障。即时健身系统也使来自不同文化背景的人们在相遇和交流时能够拥有相似的体验和情感共鸣。在这些深入的交流中，人们参与到他人的思考和创作中，体会到陪伴的作用，借此人们成为彼此的情感体验者和创造性合作者。可以说，此项目探索了艺术实践和社区关怀如何帮助人们加强人与人之间的联系，提升生活的幸福感。

设计师希望通过城市景观设计，打造城市公共空间中人与人交流的场所，服务于健身等居民的生活需求，打开人们的心扉，将关怀和爱注入到城市社区的各个角落。

杨梅竹斜街中的居民自发运动

The work "Antibodies: Landscapes in Daily Life-Accessible Fitness Systems", which features the fitness systems installed in Beijing's Yangmeizhu Xiejie, the Jing River riverbank in Beijing's sub-center, and Sheger Park in Addis Ababa, Ethiopia, was selected for the 4th China Design Exhibition and Public Art Thematic Exhibition. For this work, the View Unlimited Landscape design team considered the development of public art in the post-epidemic period from the perspective of connective-relational aesthetics.

With the emergence of the coronavirus epidemic, people's social spaces have continued to become ever more constrained. Social activities have been drastically reduced, and a decline in general health has led to increased medical costs. In response to this, accessible fitness systems provide a tangible model that can be replicated on a broad scale to promote the physical and mental health of the general population, cultivate healthy lifestyles, and improve the overall wellness of the population, while at the same time providing a feasible alternative for low-income people to gather in communities and promote the harmonious development of society. Through these intimate exchanges, people can begin to experience a sense of companionship while engaging with the intellectual and creative processes of others and gradually grow through emotional experiences and creative collaboration, by exploring how artistic practices and community engagement can offer support in difficult times.

Through the application of urban landscape design, the architect's intention is to construct a space for communication in urban public spaces, or more specifically, through the design of fitness systems and other facilities closely related to the needs of the residents, to unlock people's long-closed hearts and allow a spirit of goodwill to flow in, which can be infused throughout the entire community.

谢格尔公园的台阶上年轻人利用无器械健身系统进行科学的运动

谢格尔公园里，人们在无器械健身系统的引导下进行健身活动

（4）统筹合作优化施工方案

2017年年底，镜河水系改造工作全面展开。北京龙腾园林绿化工程公司和北京世纪经典园林绿化有限公司承接了镜河水系的建设工作。

项目对景观设计和施工团队提出了很高的要求，主要体现在多专业交叉施工和复杂的现场环境等方面。由水务部门承建的河道两侧有排水暗涵，河道末端建设排涝泵站，河道两端修建水闸，地下埋设有地源热泵、热力小室、燃气管线、地铁及综合管廊，地上密布地铁风井、地热孔、综合管廊通风井等出地面设施，核心观赏区域运河大街景观桥下方为北京地铁6号线。现场遍布各类机具、管线，且施工水电需要自行解决，各部门的施工组织急需统筹协调。在此种情况下，景观设计和施工团队坚持提前入场，统筹协调各专业部门。在满足设施使用需求的前提下，景观设计优化调整竖向种植方案，为实现"风景河道"保驾护航。

面对施工中出现的各类问题，如地下管线埋设深度不足、其他部门方案调整造成的场地条件变化等，景观设计和施工团队根据实际情况统筹协调相关各部门，同步调整景观设计方案再进行施工。例如，由于河道内水利U形槽受地铁线路影响外扩，河道与建筑红线之间的绿地面积减少，造成河道两侧地势高差加大，原方案挡墙高度无法满足现场要求。如果简单地增加挡墙高度，对河道景观效果势必造成较大影响，并且这个区域就在核心的观赏范围内。面对这一情况，景观设计团队通过点缀自然

(4) Optimizing Design Solutions through Coordinated Cooperation

By the end of 2017, the renovation of the Jing River water system was in full swing. Beijing Longteng Gardens Virescence Engineering Company and Beijing Century Classical Landscape and Gardening Co.,Ltd. undertook the construction work for the Jing River water system.

The project placed high demands on the landscape design and construction teams, particularly in terms of the project's multi-disciplinary construction and complex site environment.Drainage culverts were positioned on both sides of the river, which were constructed by the municipal water department, while drainage pumping stations and sluice gates were set up at each end of the river. Ground source heat pumps, heat chambers, gas pipelines, a subway, and a comprehensive pipeline corridor were installed underground, while the ground above was densely covered with subway wind shafts, geothermal wells, and an integrated pipeline corridor of ventilation wells and other facilities. The core scenic area, located below the landscaped bridge on Yunhe Street, was selected as the site for line 6 of the subway.The site featured all kinds of machinery and complicated pipelines and the water and electricity required throughout the construction process posed its own challenges. Moreover, prompt coordination was required between each department.Under such circumstances, the landscape design and construction teams insisted on early entry to the site and coordinated collaborative works between the various disciplines involved.Based on the premise of meeting the needs of users of the facilities, the landscape design focused on optimizing the landscape by adopting a vertical garden and planting plan to achieve a sustainable scenic riverside.

In the face of various issues which surfaced during the construction process, such as insufficient depths to install underground pipelines, changes in site conditions caused by other departments' adjustments, etc., the landscape design and construction teams coordinated with all of the relevant departments in accordance with the situation on the ground and adapted the landscape design plan concordantly prior to commencing construction.For example, due to the subway line's outward expansion and its effect on the waterway's U-channel, the green areas between the river and the building site's red lines were reduced, resulting in greater variations in terrain height on both sides of the river, as the original scheme for the retaining wall's height could not meet the site requirements.If the height of the retaining wall were simply raised, it would have inevitably had a greater impact on the riverscape and detract from the core scenic

景石的方式处理地形高差，突出景观节点，避免采用过于人工的方式处理。施工过程中遇到类似问题，景观设计团队会从景观桥、观景平台、路网等多个视角检验施工效果，做到移步异景，实现自然景观的丰富变化。

镜河河道两岸现状标高低于设计标高，回填土作业需要机械施工，但河道两侧绿地内各有一条水利暗涵，机械施工难度很大，限制了施工时土壤回填的密实度，从而影响绿化种植层的稳定性。根据这一情况，在采用传统机械压实、水夯处理的基础上，景观设计团队及时调整植物品种，选择根系发达的植物，如银杏、油松、白皮松、元宝枫、国槐、沙地柏、迎春等，解决了回填土密实度不足的问题。而针对回填土成分复杂、肥力不足的情况，施工团队做好土壤改良措施，多次送检土样，提升土壤理化性质，使其适于植物生长。

景观设计和施工方协同合作，严格把控苗木质量，合理修剪苗木，确保苗木满足设计效果。本工程的施工贯穿冬、春、夏、秋四个季节，在市园林绿化局专家的指导下，合理运用修剪技术，在确保成活的前提下立地成景；采用多种种植方式和搭配方法，应对不同立地条件，加强草花组合、观赏草与宿根花卉等植物的综合运用；利用植物材料枝、杆、叶的绿期对铺装、绿地等细节部位进行修饰，在实现建设目标及景观效果的同时，创造稳定、健康的植物群落结构，创建舒适宜人的河道环境。

area.Faced with this situation, the landscape design team approached the disparity in terrain height by embellishing the site with natural scenic rocks and highlighted the landscape's nodes while avoiding an overly artificial solution.When encountering similar problems during construction, the landscape design team examined the construction's effect from multiple perspectives, such as the landscape's bridges, viewing platforms, and network of paths, to achieve a rich and varied natural landscape which evolved progressively from one scene to the next.

The elevation of both sides of the Jing River riverbank was lower than the design intended, and backfilling operations required mechanical construction. The green spaces on both sides of the river featured culverts for water conservation, which made mechanical construction difficult and limited the compactness of the backfilled soil, thus affecting the sustainability of the site's greening and planting.In light of this situation, and on the basis of traditional mechanical compaction and rammed earth treatment, the landscape design team pivoted quickly and adjusted the site's plant species by choosing plants with developed root systems, such as *Ginkgo biloba, Pinus tabuliformis, Pinus bungeana, Acer truncatum, Sophora japonica, Sabina vulgaris* and *Jasminum nudiflorum*, so as to address the issues of insufficient backfilling and compactness.As for tackling the complex composition and poor fertility of the backfilled soil, the construction team implemented measures to optimize the soil and sent samples several times to improve the physical and chemical properties of the soil to make it more suitable for planting.

The landscape design and construction teams worked together to strictly control the quality of seedlings and applied appropriate pruning to ensure that the seedlings achieved the intended effect of the design team.The construction for the project extended through the four seasons and under the guidance of experts from the Municipal Bureau of Landscaping, suitable pruning techniques were used throughout this period to ensure the integrity of the landscape. A variety of planting methods and arrangements were used to cope with the various site conditions, while the integrated use of plants, grass combinations, ornamental grasses and perennial flower enhanced the overall effect. By utilizing the "branches, stems, leaves and green phases" of plants to embellish the pavement, green areas and other related aspects, the team realized the construction goals and overall effect of the landscape, while creating a sustainable and healthy plant ecosystem and a more comfortable and inviting riverside environment.

（5）人地和谐促进绿色发展

镜河河道景观设计坚持绿色发展理念，在中国传统自然观"天人合一"思想的指导下，用诗意化的人工手法因地制宜地弥补自然缺陷，既营造了河道自然环境，又展现出中国自然山水画的意境。在河道外迁、水岸环境塑造的过程中，努力打破人与自然、人工与原生态间的界限，充分利用已有资源（地形地貌、林木石材等），富于创意地借鉴中国传统园林艺术的"借景"、传统绘画艺术的"留白"，尽可能减少不必要的扰动，避免强行介入，以"宜人"为目的实现环境的改善，创造良好而丰富的景观与生态系统。

(5) Harmonious Green Development Between People and Nature

The landscape design of the Jing River adheres to the concept of green development, which is imbued with the traditional Chinese concept of nature and "the unity of man and heaven", and uses poetic man-made techniques to remedy shortcomings in the natural environment in accordance with local conditions, so as to not only create a natural environment for the river, but also showcase the artistic temperament of Chinese natural landscape paintings. In the process of relocating the river and shaping the waterfront, the team strove to break the boundaries between humans and nature, and between the original ecology and man-made structures. By making full use of existing resources (the present topography and geomorphology, forest and stones, etc.), the team creatively drew from the concepts of "borrowed scenery" in traditional Chinese garden art and "white space" in traditional paintings in order to reduce unnecessary disturbances to the greatest extent possible, avoid forcible intervention, achieve a better and more pleasant environment, and create a stunning and diverse landscape and ecosystem.

The design team focused on guiding the co-construction of shared green space and inspiring people to get closer to nature and to care for the environment through experiential features and participatory activities, so as to promote

通过设计衔接人、建筑、人工景观与自然之间的关系,展现中国传统自然观

设计团队注重引导绿色空间的共建共享，通过可体验的场景、可参与的活动，引导人们亲近自然、爱护环境，促进人地关系的协调发展，促进绿色与可持续发展理念的传承。同时，增强副中心办公人群及周边居民的归属感，增进人与人之间的交流互动，为城市新区发展积聚人气与能量。

建成的镜河风景河道对保护场地生态、提升生态水平，以及稳定生物群落等方面，起到了较为显著的促进作用。

harmonious development between humans and the earth, and further encourage the continuity of green and sustainable development concepts. At the same time, the design enhances a sense of belonging in the hearts of the sub-center's office population and surrounding residents, while improving social exchanges and interactions between people and bolstering the popularity and dynamism of the new urban area and its future development.

The construction of the Jing River riverscape has contributed significantly to the conservation and landscape ecology of the site, as well as to the stabilization of the biome.

片区内对石料的应用

资料链接6

"都市花草堂"共建——公共艺术活动策划

"都市花草堂"源于无界景观工作室2015年在北京杨梅竹斜街环境更新与公共空间营造项目中发起的"胡同花草堂"活动。

在杨梅竹斜街项目的前期调研中,设计团队发现了种植的另一种形式:胡同中的住家大多喜欢在自家周边狭窄的空间缝隙,以随手即拾的容器种植自己喜欢的花草和日常食用的蔬菜。这种自家种植的花草与蔬菜在外观上也许不如公共绿化的植物美观,种植器皿也可能五花八门、粗糙简陋,但人们对于充满生活趣味的花开花落、植物生长的喜好,并没有因物质条件的简陋而减少。这应该是一种心灵的慰藉,一种现实生活中的精神寄托。

Link 6

Urban Flora Cottage Co-construction and Planning

The Urban Flora Cottage originated from the Hutong Flora Cottage project, which was initiated by View Unlimited Landscape Architecture Studio in 2015 as a part of the environmental renewal of Beijing's Yangmeizhu Xiejie and creation of public space.

During the preliminary survey of the Yangmeizhu Xiejie project, the design team discovered another social function of planting: people generally preferred to grow their favorite flowers and vegetables for daily consumption in containers that they could pick up easily in the narrow spaces between their residences. Although such home-grown flowers and vegetables were not as aesthetically pleasing as the plants used in public greening, and their planting utensils were varied and crude in nature, the people's interest in flowers and plants had grown to become an important part of their daily life and was not diminished by their limited material conditions. This act essentially functioned as a sort of emotional solace, or spiritual anchor in the midst of their reality.

杨梅竹斜街"胡同花草堂"项目《安住·平民花园》入选威尼斯建筑双年展

无界景观工作室以建立"胡同花草堂"的方式,将"种植"这种自发的"个体行为"转化为"公共行为",鼓励胡同居民参与社区环境改造,通过养花、种菜等形式拉近相互间的距离,美化街道环境,增强居民的主体意识,传递邻里间的守望相助与温暖包容。在这个过程中,设计师运用专业知识为社区建立有机结构,激发原住居民的创造力与生活热情——设计师的角色转变了,设计工作从"为居民设计"转变为"引导居民自发营造居住环境"。

　　2016年,杨梅竹斜街"胡同花草堂"项目入选威尼斯建筑双年展,杨梅竹斜街的居民通过视频与来自世界各地的参观者互动,与国际舞台的连接使居民进一步增加了自信心与荣誉感。2019年,杨梅竹斜街"胡同花草堂"公共艺术作品入选第三届中国设计大展及公共艺术专题展。

　　在北京城市副中心行政办公区镜河景观设计项目中,无界设计团队曾策划以镜河河岸的生态挡墙为载体,组织办公人群和周边居民在挡墙内种植乡土攀援植物及草籽组合,通过这些植物隐藏人工痕迹,使墙体掩映于自然之中。遗憾的是,镜河"都市花草堂"共建公共艺术活动因多种原因未能实现。

　　2020年,无界景观工作室把"都市花草堂"活动策划移植到埃塞俄比亚首都亚的斯亚贝巴城市中心的谢格尔公园中。设计团队在现场指导园林施工,向埃塞俄比亚传递中国园林文化和科技创新技术的同时,策划组织埃塞俄比亚政府官员、市民代表以及中方工作人员在公园长达千米的挡土墙上种植花草,"共谋绿色生活,共建美丽家园"。目前这座"花墙"已初具规模,这将是一个

2016年威尼斯双年展的参观者通过视频与杨梅竹斜街的居民互动

2021年北京国际设计周期间杨梅竹斜街的居民与游客参加了中埃两国人民的"花草堂"启动仪式并交流两国文化

By establishing the "Hutong Flora Cottage", View Unlimited transformed the spontaneous and individual behavior of planting into a public activity that encouraged hutong residents to participate in the transformation of their communal environment, thus bringing residents closer to each other through natural intermediaries such as cultivating flowers and planting vegetables, beautifying the surrounding environment, strengthening a feeling of ownership among residents, and imparting a greater sense of mutual care, warmth, and tolerance. Throughout this process, our designers used their expertise to create an organic structure for the community and to stimulate the creativity and enthusiasm of local residents, thus transforming the designers' role from "designing for residents" to "guiding residents in creating their own living environment".

In 2016, Yangmeizhu Xiejie's "Hutong Flora Cottage" project was selected for the Venice Biennale of Architecture, which provided a platform for the residents of Yangmeizhu Xiejie to interact with visitors from all over the world via a live video stream and further boost their confidence and sense of pride by connecting with an international stage. In 2019, the "Hutong Flora Cottage" was subsequently selected as a public art work for the Third China Design Exhibition and Public Art Thematic Exhibition.

As part of the landscape design of the Jing River, which runs along the administrative offices of Beijing's sub-center, the View Unlimited design team planned to use the ecological retaining wall on the Jing River riverbank as a medium to rally office workers and neighboring residents to plant native climbing plants and grass seed combinations inside the wall, so as to hide any man-made traces of the design through natural plant growth and conceal the wall in nature. Unfortunately, the "Urban Flora Cottage" public art event on the Jing River could not be implemented due to various circumstances.

In 2020, View Unlimited Landscape Studio transplanted the "Urban Flora Cottage" project to Sheger Park, in the heart of Addis Ababa, Ethiopia. The design team, who were

持续的计划，通过后续的不断完善，最终这个千米长的挡土墙会变成一座美丽的"岩石花园"。设计师不会规定种什么花，能生长就行，当地民众可以发挥创造力，植物会越长越丰富。这是景观设计师和当地民众共同参与的活动，是一个大型公共艺术计划。该种植计划还被拓展至当地的棚户区，教授居民在有限的空间中实现蔬菜立体种植，既美化了环境又提供了食物来源。

2020年1月1日，在谢格尔公园友谊广场媒体开放日期间举办的"都市花草堂"活动得到埃塞俄比亚阿比·艾哈迈德总理的大力支持并亲自参与。2021年北京国际设计周期间，一场联结中埃两国人民的"花草堂"启动仪式引发各界关注。启动仪式在埃塞俄比亚首都亚的斯亚贝巴友谊广场和北京杨梅竹斜街同时开启，通过网络直播连线，两国居民进行互动和文化交流：举办"花草堂"挂牌和播种仪式，观赏埃塞俄比亚独特的咖啡仪式、服饰演出和中国文化主题表演，包括采瓷坊金砖拓印传统活动等。透过大屏幕，两国人民感受着彼此的异域世界，相互欣赏，相互问候，花草堂活动成为中国和埃塞俄比亚"一带一路"共建共享、民心相通的创新实践。

在线直播埃塞俄比亚独特的咖啡仪式、服饰演出，促进两国人民友谊与文化交流

on site to instruct the garden's construction, focused on conveying both the tradition of Chinese garden culture and scientific and technological innovation, and planned and organized Ethiopian government officials, civic representatives, and Chinese staff in planting flowers and plants along the park's kilometer-long retaining wall in the spirit of "seeking a greener life and creating a beautiful home together". This flower wall is now taking shape and will continue as an ongoing project that, hopefully, can be further improved through subsequent organized activities, until eventually, the retaining wall will be transformed into a beautiful rock garden. Rather than dictating which flowers and plants to choose, the designers decided to encourage the use of species which grow naturally, so that residents can unleash their creativity and let their creations flourish. The initiative, which represents a joint activity between the team's landscape architects and residents, is to be carried out as a large-scale public art project. Moreover, the program has also been extended to local slums, where residents can learn to grow vegetables in constricted spaces by applying three-dimensional models, thus beautifying the environment, and addressing some of the daily challenges they face.

在埃塞俄比亚通过网络直播连线，两国人民进行互动和文化交流

On January 1, 2020, the "Urban Flora Cottage" event was held during a media day hosted in Sheger Park's Friendship Square, with the support and participation of Ethiopian Prime Minister Abiy Ahmed. Beijing Design Week 2021 marked the launch of "Flora Cottage", an event linking together the people of Ethiopia and China and garnered the attention of people from all walks of life, with the opening ceremony taking place simultaneously in Friendship Square in Addis Ababa, Ethiopia, and Yangmeizhu Xiejie in Beijing. Via a live webcast, residents of the two countries interacted and engaged in cultural exchanges, including an official plaque, seeding ceremony for "Flora Cottage", special Ethiopian coffee ceremony, performance featuring Chinese traditional costumes and culturally themed show, and the traditional activity of gold tile rubbings held in the Porcelain Workshop. Connected across a large screen, people experienced novel and unique elements from their two respective worlds. As participants gazed upon and greeted each other, the Flora Cottage event served as an innovative model that reflected the spirit of the Belt and Road Initiative between China and Ethiopia, creating a platform to connect and bring two peoples closer together.

谢格尔公园千米长的挡土墙成为一座美丽的"岩石花园"

资料链接7

北京副中心行政办公区先行启动区生物多样性调查

北京城市副中心行政办公区先行启动区内种植的植物品种多达130余种，项目建成并投入使用后，生物爱好者李同宇先生于2021年夏天对数十种植物和近30种动物进行了为期一天的生物多样性调查。根据首都文明网、潇湘晨报、京报网等媒体报道，副中心记录的野生鸟类已达353种（含历史记录21种），其中仅镜河周边鸟类已记录19种，包括斑嘴鸭、棕背伯劳、黑水鸡、红隼等，其中不乏珍稀鸟类，鱼类15种。这些动植物已经构成了较为完善的生境、食物链、食物网、生态链等，为生物多样性提供基础。

生物多样性调查研究中主要物种汇总

乔木				
序号	名称	拉丁名称	科属	特点
1	栾树	*Koelreuteria paniculata*	无患子科 栾树属	观赏性强，能提供栖息场所
2	垂柳	*Salix babylonica*	杨柳科 柳属	可遮阴，观赏性强，能提供栖息场所
3	油松	*Pinus tabuliformis*	松科 松属	常绿植物，观赏性强，能提供栖息场所
4	七叶树	*Aesculus chinensis*	七叶树科 七叶树属	观叶、观花、观果，观赏性强
5	彩叶豆梨	*Pyrus calleryana*	蔷薇科 梨属	秋季季相变化，观赏性强
6	皂荚树	*Gleditsia sinensis*	豆科 皂荚属	可入药，可当肥皂，寓意好
7	白蜡	*Fraxinus chinensis*	木犀科 白蜡属	秋季季相变化，观赏性强，有药用价值
8	碧桃	*Amygdalus persica*	蔷薇科 桃属	春季开花，观赏性强
9	山桃	*Amygdalus davidiana*	蔷薇科 桃属	3~4月开花，观赏性强
10	紫荆	*Cercis chinensis*	豆科 紫荆属	3~4月开花，观赏性强
11	黄栌	*Cotinus coggygria*	漆树科 黄栌属	秋季季相变化，观赏性强，有药用价值
12	加拿大红樱	*Prunus virginiana*	蔷薇科 李属	叶常年紫红色
13	山杏	*Armeniaca sibirica*	蔷薇科 杏属	沙荒防护林的伴生树种，经济价值高
14	紫薇	*Lagerstroemia indica*	千屈菜科 紫薇属	夏秋开花，花期长，观赏性强
15	柽柳	*Tamarix chinensis*	柽柳科 柽柳属	枝条细柔，姿态婆娑，观赏性强
16	龙柏	*Sabina chinensis*	柏科 圆柏属	常绿植物，枝条螺旋盘曲，观赏性强
17	元宝槭	*Acer truncatum*	槭树科 槭属	秋季季相植物，观赏性强
18	榆叶梅	*Amygdalus triloba*	蔷薇科 桃属	开花，观赏性强
灌木				
序号	名称	拉丁名称	科属	特点
1	迎春	*Jasminum nudiflorum*	木犀科 素馨属	春季开花，适应性强
2	沙地柏	*Sabina vulgaris*	柏科 圆柏属	常绿植物，匍匐灌木
3	蔷薇	*Rosa sp.*	蔷薇科 蔷薇属	开花，观赏性强
4	平枝栒子	*Cotoneaster horizontalis*	蔷薇科 栒子属	枝密叶小，红果艳丽
5	欧洲琼花	*Viburnum opulus*	忍冬科 荚蒾属	花大密集，叶秋季变红
6	金银木	*Lonicera maackii*	忍冬科 忍冬属	春季金银花色，秋冬季观果
7	紫薇	*Lagerstroemia indica*	千屈菜科 紫薇属	夏秋开花，花期长，观赏性强
8	紫荆	*Cercis chinensis*	豆科 紫荆属	3~4月开花，观赏性强
9	连翘	*Forsythia suspensa*	木犀科 连翘属	黄色花，观赏性强
10	丁香	*Syringa oblata*	木犀科 丁香属	春季开花，芳香植物，观赏性强

续表

灌木

序号	名称	拉丁名称	科属	特点
11	黄刺玫	*Rosa xanthina*	蔷薇科 蔷薇属	黄白色花，观赏性强
12	棣棠	*Kerria japonica*	蔷薇科 棣棠花属	5~6月开花，绿篱植物
13	大叶黄杨	*Buxus megistophylla*	黄杨科 黄杨属	绿篱灌木，常绿

地被植物

序号	名称	拉丁名称	科属	特点
1	金娃娃萱草	*Hemerocallis fulva*	百合科 萱草属	观赏性强，耐热抗寒，适应性强
2	月季	*Rosa chinensis*	蔷薇科 蔷薇属	自然花期4~9月，观赏性强
3	大花萱草	*Hemerocallis middendofii*	百合科 萱草属	观赏性强，适应性强
4	紫萼	*Hosta ventricosa*	百合科 玉簪属	开花，观赏性强
5	鸢尾	*Iris tectorum*	鸢尾科 鸢尾属	秋季季相变化，观赏性强
6	芍药	*Paeonia lactiflora*	芍药科 芍药属	开花，观赏性强，有入药价值
7	蛇莓	*Duchesnea indica*	蔷薇科 蛇莓属	开花结果，观赏性强，有入药价值
8	扶芳藤	*Euonymus fortunei*	卫矛科 卫矛属	地面覆盖植物，增加绿意
9	车轴草	*Trifolium*	豆科 车轴草属	绿色期长，花期长，耐践踏

水生植物

序号	名称	拉丁名称	科属	特点
1	荷花	*Nelumbo sp.*	莲科 莲属	观赏性强，生态植物
2	醉鱼草	*Buddleja lindleyana*	马钱科 醉鱼草属	观赏性强，吸引各种昆虫
3	菖蒲	*Acorus calamus*	天南星科 菖蒲属	挺水植物，开花，观赏性强
4	香蒲	*Typha orientalis*	香蒲科 香蒲属	挺水植物，观赏性强
5	水葱	*Scirpus validus*	莎草科 蔍草属	挺水植物，可为昆虫及鸟类提供栖息场所
6	芦苇	*Phragmites australis*	禾本科 芦苇属	挺水植物，观赏性强，可为昆虫及鸟类提供栖息场所
7	梭鱼草	*Pontederia cordata*	雨久花科 梭鱼草属	花色迷人，花期较长，观赏性强

昆虫

序号	名称	拉丁名称	科属	习性特点
1	菜粉蝶	*Pieris rapae*	粉蝶科 菜粉蝶属	嗜食十字花科植物，及其他寄主植物 相关场地植物：扶芳藤
2	水黾	*Aquarium paludum*	黾蝽科 大黾蝽属	以落入水中的小虫体液、死鱼体或昆虫为食。雌虫在水表面附近植物上产卵，喜将卵产在小而细的水草茎秆上，常以丝状物包蔽 相关场地植物：挺水植物、水草
3	异色多纹蜻（同色型）	*Deielia phaon*	蜻科 多纹蜻属	成虫6~9月，选择河底淤泥肥沃，有机质含量高的静水潭附近栖息，喜成群发生，栖息于挺水植物上 相关场地植物：芦苇、水葱、香蒲、荷花
4	异色多纹蜻（异色型）	*Deielia phaon*	蜻科 多纹蜻属	成虫6~9月，选择河底淤泥肥沃，有机质含量高的静水潭附近栖息，喜成群发生，栖息于挺水植物上 相关场地植物：芦苇、水葱、香蒲、荷花
5	白尾灰蜻	*Orthetrum albistylum*	蜻科 灰蜻属	成虫4~10月，栖息于挺水植物，喜欢在特定区域迂回飞行，常停歇于地面上
6	异色灰蜻	*Orthetrum melania Sely*	蜻科 灰蜻属	成虫6~8月，栖息于挺水植物，喜欢在植被或石头上休息 相关场地植物：荷花、挺水植物

续表

昆虫					
序号	名称	拉丁名称	科属	习性特点	
7	玉带蜻	*Pseudothemis zonata*	蜻科 玉带蜻属	成虫6~10月，栖息于水下有机质较多的水系，喜挺水植物、荷花等植物附近 相关场地植物：荷花、香蒲、芦苇	
8	捷尾蟌	*Paracercion v-nigrum*	蟌科 尾蟌属	成虫5~8月，栖息于植物较多的静水区域 相关场地植物：荷花、芦苇、香蒲	
9	长叶异痣蟌	*Ischnura elegans*	蟌科 异痣蟌属	成虫5~10月，栖息于静水植物茂盛环境 相关场地植物：扶芳藤、荷花、香蒲、芦苇	
10	苇尾蟌	*Paracercion calamorum*	蟌科 尾蟌属	成虫6~8月，栖息于植被茂盛、水草较多及挺水植物丰富的静水区域。停歇在挺水植物上 相关场地植物：荷花、香蒲、芦苇	
11	联纹小叶春蜓	*Gomphidia confluens*	春蜓科 小叶春蜓	成虫5~8月，在挺水植物上休息，具有领地意识 相关场地植物：荷花、香蒲、芦苇	
12	碧伟蜓	*Anax Parthenope*	蜓科 伟蜓属	成虫5~9月，捕食各类昆虫，交配后在水草、枯木产卵，稚虫捕食小鱼小虾 相关场地植物：荷花、香蒲、芦苇	
13	闪蓝丽大伪蜻	*Epophthalmia elegans*	伪蜻科 丽大伪蜻属	成虫5~9月，在水系绕圈飞行 相关场地植物：荷花、香蒲、芦苇	
鸟类					
序号	名称	拉丁名称	科属	习性特点	
1	麻雀	*Passer*	雀科 麻雀属	多在有人类集居的地方活动和觅食。杂食性，以禾本科植物种子、昆虫、鳞翅目害虫、农田谷物等为食	
2	小䴘䴘	*Tachybaptus ruficollis*	䴘䴘科 小䴘䴘属	善游泳潜水，以水生昆虫、鱼、虾等为食。繁殖时在水上追逐，在丛生的芦苇、灯心草、香蒲等地营巢	
3	黑水鸡	*Gallinula chloropus*	秧鸡科 黑水鸡属	栖息于芦苇和挺水植物较多的水系。以水生植物及水生昆虫、蠕虫、蜘蛛、软体动物、蜗牛等为食	
4	赤膀鸭	*Anas strepera*	鸭科 鸭属	栖息和活动在内陆水域中，尤喜在水生植物较多的开阔水域活动。以水生植物为食，常在水边水草丛中觅食	
5	白头鹎	*Pycnonotus sinensis*	鹎科 鹎属	结群于果树上活动。多活动于树木灌丛中，也见于针叶林里。吃大量的农林业害虫，农林益鸟之一，值得保护	
6	喜鹊	*Pica pica*	鸦科 鹊属	喜欢将巢筑在民宅旁的大树上。以昆虫、蛙类等小型动物为食，兼食瓜果、谷物、植物种子等	
7	灰喜鹊	*Cyanopica cyanus*	鸦科 灰喜鹊属	栖息于开阔的松林及阔叶林，公园和城镇居民区。以半翅目、鞘翅目的昆虫及幼虫，植物果实为食	
8	大杜鹃	*Cuculus canorus*	杜鹃科 杜鹃属	栖息于开阔林地，特别喜欢近水的地方。平时仅听到鸣声，很少见到。以鳞翅目幼虫、甲虫、蜘蛛、螺类等为食。食量大，对消除害虫起相当作用	
9	夜鹭	*Nycticorax nycticorax*	鹭科 夜鹭属	白天结群隐藏于密林中僻静处，夜出性。喜结群。主要以鱼、蛙、虾、水生昆虫等动物性食物为食	
10	家燕	*Hirundo rustica*	燕科 燕属属	停落在树枝、电杆和电线上，飞行时张着嘴捕食蝇、蚊等各种昆虫。鸣声尖锐而短促	
11	黄苇鳽	*Ixobrychus sinensis*	鹭科 苇鳽属	喜栖息在既有开阔水面又有大片芦苇和蒲草等挺水植物的地方。以小鱼、虾、蛙、水生昆虫等为食。繁殖期为5~7月，营巢于浅水处芦苇丛和蒲草丛中	
12	白鹡鸰	*Motacilla alba*	鹡鸰科 鹡鸰属	栖息于离水较近的地方。经常成对活动或结小群活动。以昆虫为食。觅食时地上行走，或在空中捕食昆虫。	
鱼类					
序号	名称	拉丁名称	科属	习性	
	窜条鱼/餐条鱼	*Hemiculter Leucisclus*	鲤科 鲹属	行动迅速，常群游于浅水区上层。杂食，主食无脊椎动物。分批产卵，黏附于水草或砾石上	

· 生境构建

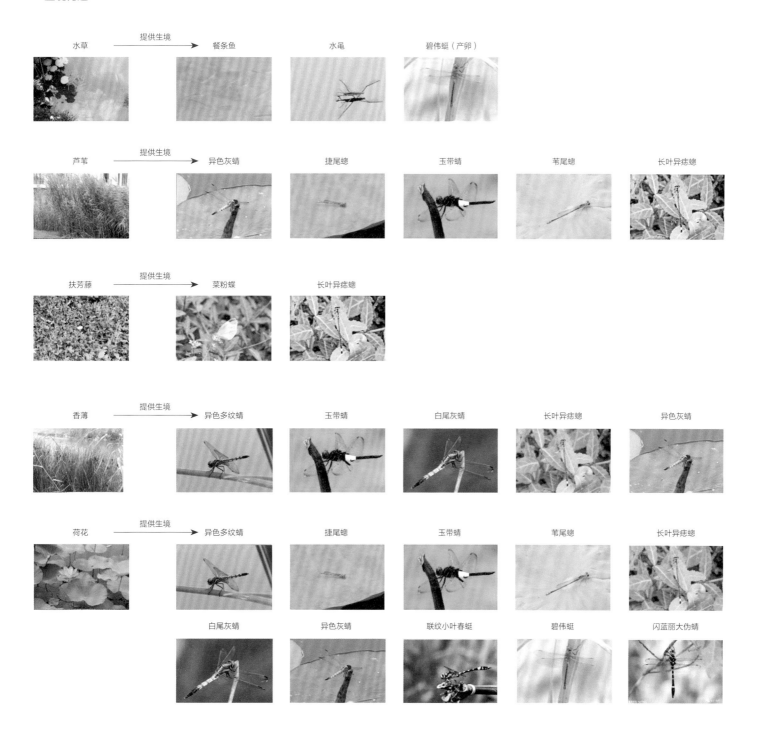

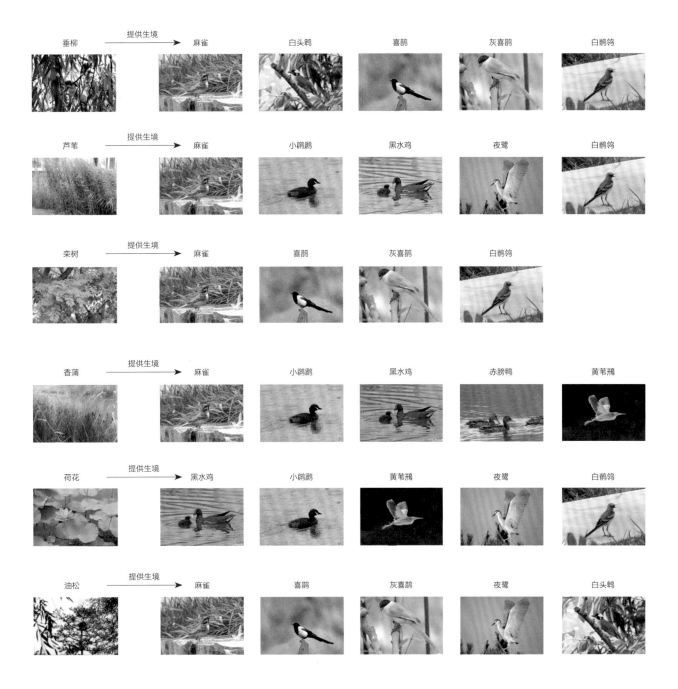

蜻蜓被誉为"水质检测器",因为蜻蜓稚虫生活在水中,水质更好的区域会拥有更丰富的蜻蜓资源。异色多纹蜻和玉带蜻有"淤泥肥沃、有机含量高"的生境需求,因此场地内有一定数量的异色多纹蜻和玉带蜻代表了场地水质良好,淤泥肥沃。

不同的水鸟有不同的生活习性,但基本都需要水生植物丰富、水质良好的栖息环境。场地内的不同植物可以提供给多种鸟类生存空间。如小䴘喜欢生存在芦苇及香蒲的环境中,黑水鸡栖息在芦苇丛中或荷花池,黄苇鳽喜大片芦苇和蒲草的环境等。

· 食物网构建

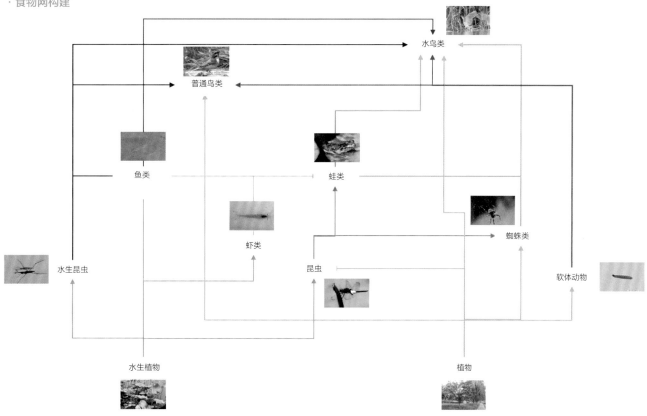

较为宏观概括的食物链骨架营建

· 生态链构建

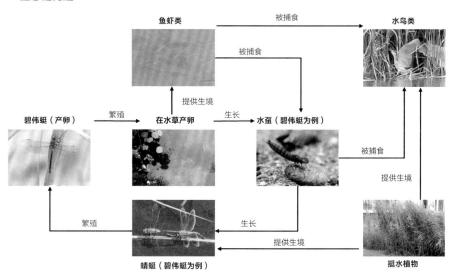

植物、昆虫、鸟类之间相互联系，共同构成生态链网结构。丰富的自然空间可以提供给动植物更好的生境场所，丰富的动植物资源也可以反映出当地生态的良好程度。调查区域拥有丰富的动植物资源、科学生态的景观空间，具有良好的生态价值。

· 场地生态实景

A. 岸线植物配置
植物种类：丁香+迎春+荷花+醉鱼草；主要功能：植物芳香、观赏视角、昆虫生境。

B. 萱草景观节点
植物种类：大花萱草+栾树+垂柳+荷花；主要功能：鸟类栖息、人群游览、昆虫生境。

C. 观景平台
植物种类：金娃娃萱草+丁香+垂柳+荷花+芦苇；主要功能：观景平台、鸟类栖息、昆虫生境。

D.廊架前道路

植物种类：鸡爪槭＋榆叶梅＋荷花＋芦苇＋车轴草；主要功能：鸟类栖息、人群游览、昆虫生境。

E.健身步道

植物种类：黄栌＋丁香＋彩叶豆梨＋垂柳；主要功能：有益身心、健身步道、动物生境。

F.沿步道休憩座椅

植物种类：旱柳＋锦带花＋蔷薇＋香蒲＋鸢尾；主要功能：健身步道、动物生境、休憩观景。

天津渤龙湖经济区综合体景观设计

LANDSCAPE DESIGN OF
THE TIANJIN BOLONG LAKE
ECONOMIC ZONE COMPLEX

天津渤龙湖经济区综合体景观设计
Landscape Design of the Tianjin Bolong Lake Economic Zone Complex

项目地点： 中国 天津市滨海高新科技园区	**Project location:** Tianjin, China
项目规模： 60 公顷	**Project scale:** 60 hectare
设计时间： 2009—2011 年	**Design period:** 2009-2011

项目背景

天津渤龙湖经济区综合体位于天津滨海高新科技园区，距离天津市中心30公里，是一个集总部写字楼、研发中心、星级酒店、商业商务中心、高档住宅区和滨水生态景观于一体的综合性经济区域，未来将以"新"为追求，引入新商业与新经济。本项目是无界景观工作室与多家设计单位❶共同参与的集群设计，集合了不同设计师的思想和手法，具有多样性和复杂性，并具有前瞻性和实验性。

项目的位置和性质决定了此集群设计的目的是产生某种"合力"，能够获得规模效应。聚集的优势不仅在于得到社会与媒体的关注，更在于促进交流，促进创新，构建虚拟与现实的双重网络，实现"云时代"资源的优化、整合与共享❷。

天津渤龙湖经济区综合体区位

---

❶ 建筑设计单位包括：齐欣建筑、天津华汇工程建筑设计有限公司、中国建筑设计研究院李兴刚工作室、清华建筑设计院庄惟敏工作室、北京市建筑设计研究院王戈工作室。

❷ 参考资料：《中国建筑艺术年鉴》(2011—2012年)。

设计理念

户外环境是人们交流交往的平台，能够实现人与环境、人与人之间的连接。人们在此享受不同思维碰撞带来的智慧和乐趣，汲取自然赋予人们的能量，提升生活品质，促进社会的和谐发展。因此，景观设计团队将"乐活景观（Lohascape）"的概念融入其中，在整体景观设计中体现"乐活"理念，通过设计舒适的户外环境，改变人们以往的行为与生活方式，鼓励人们走向户外，进行健身、步行、骑行等户外活动。个人行为方式的转变将有效降低碳排放，达到保护自然、改善居住环境的目的。环境设计引发的人类行为与生活方式的转变符合园区"新"的特征，并将通过集群模式放大，通过"合力效应"增强对自然、社会、文化的影响。在面对人口与资源双重压力的今天，倡导新的可持续的生活方式和新的消费观念必然成为一种趋势，对建设环境友好与资源节约型社会也具有非常重要的意义。

乐活景观（Lohascape）理念

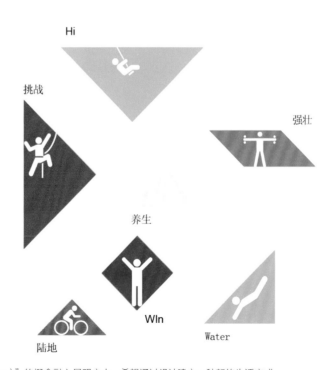

将"乐活景观（Lohascape）"的概念融入景观之中，希望通过设计建立一种新的生活方式

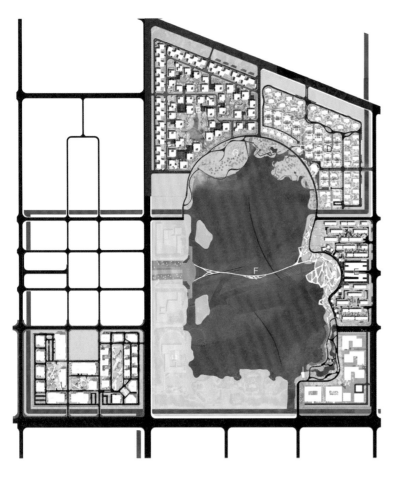

A 总部基地一区
B 总部基地二区
C 总部基地三区
D 生态居住一区
E 生态居住二、三区
F 跨湖步行栈桥
G 环湖绿化带及步行道

天津渤龙湖经济区综合体总平面图

在具体设计手法上，由于整个区块涵盖不同的功能属性，同时由不同的设计事务所完成各地块的建筑设计，因此考虑通过一个完整体系对园区环境资源进行整合，应用景观设计的方式将建筑与建筑、建筑与环境紧密联系在一起，使道路、各类活动场地、构筑物、水景、植物等景观元素共同构成连续而美好的景观形象。

总部基地区的景观设计延续渤龙湖区块景观肌理，结合功能化绿地创造出丰富多变、富有诗意的景观场景，鼓励人们走出办公室，在此交流办公、放松身心，提倡健康的工作、生活方式。生态居住区的景观设计在保持区域景观连续性的同时，叠加居住区自身的景观特色，院落景观结合镜面水池、花溪水巷以及引入的湖岸线，形成舒适宜人的四季自然美景；高层建筑则结合公园式的开放景观，高大树木能有效消减高层给人们带来的压抑感觉。

设计团队将"穿行"的理念融入其中，希望人们在穿行的过程中体会生活的美好。在现有车行路基础上，叠加步行网络，人们可以轻松自如地从建筑内部进入自然环境，或从一个区域到达另一个区域。步行道路将一个个美好的景观片段连缀起来，强调戏剧化的步行体验——视角的高低变化、四季的更替、天气的阴晴、植物的姿态。步行成为人们每日生活必不可少的内容，人们享受其中，边走边玩，生活节奏也随之放慢。在渤龙湖上，步行道路成为一座栈桥，建立了东西两岸的联系，栈桥具有丰富的形态变化，并在湖心形成廊架和活动场地，为人们带来丰富的空间体验和不同的观湖视角。

"穿行"手法的最终目的是营建小尺度、高密度、亲近人的邻里环境，成为人们进行交往、休闲、娱乐等活动的城市中最具特色和最具感染力的公共活动空间，并以此提升居住者的归属感与认同感。

"穿行"的公园将周边多样化的功能贯穿到一起，形成一个尺度亲密、能激发各种互动和活力的公共空间领域，能行使"城市"功能的综合体。各种流动的多样化的空间为丰富多彩的城市生活和活动提供展示的舞台，满足多样性的使用需求。这种高复杂度的新的城市空间形态能作为一种弹性体系适应未来城市变化发展的可能。

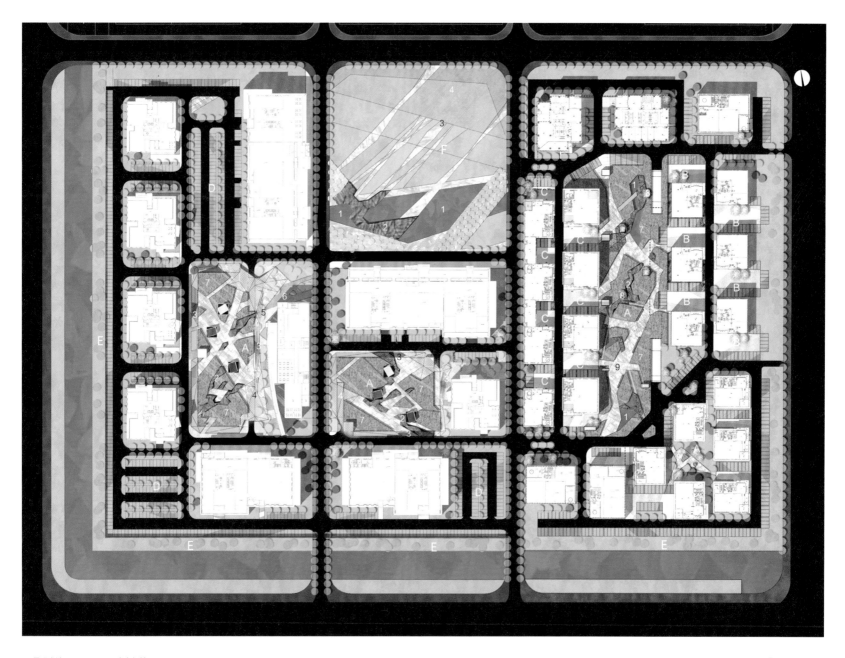

1 景观水池
2 景观构筑物
3 景观台阶
4 景观折坡
5 银杏
6 石材铺装
7 竹林
8 地下车库树池
9 嵌草铺装

A 中心绿地
B 建筑庭院一
C 建筑庭院二
D 绿化停车场
E 景观绿带
F 区内小游园

天津渤龙湖经济区综合体总部三区平面图

天津渤龙湖经济区综合体鸟瞰图

天津渤龙湖经济区综合体跨湖桥鸟瞰图

总部基地二区中心湖

总部基地二区中心湖夏景

生态居住一区剖面图

建筑　　微地形　　公园式大道　　微地形　　停车场

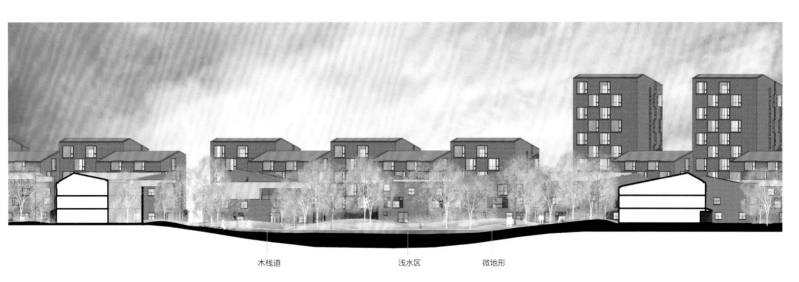

木栈道　　浅水区　　微地形

总部基地二区剖面图

生态居住一区
公园式大道景观

生态居住一区夏季景观(一)

生态居住一区
小高层景观

生态居住一区夏季景观（二）

天津解放南路地区景观规划及公园设计

LANDSCAPE PLANNING AND PARK DESIGN OF SOUTH JIEFANG ROAD, TIANJIN

项目背景

天津是依托海河发展起来的移民城市和商埠城市，是华北地区民族工业的摇篮。受近代文化影响，天津人眼界开阔、思想开放、善于找乐，具有十足的"乐活"精神，形成了独具特色的地域文化性格。近年来，天津工业转型升级，重心向东南转移，中心城区逐渐转变成以商贸、居住、文化功能为主的生态环境良好的宜居城市。

位于天津市中心城区南部的解放南路地区，处于天津市发展主轴线上、天津主副城市中心之间，北接海河，南邻天津外环路，是海河沿线的重要发展地区。地块内原有大片工业区，工业重心转移后，区块内保留的工业遗址共有四处，分别是天津市公共交通二公司、天津市电机总厂、天津市渤海无线电厂及天津市陈塘庄热电厂，它们见证了天津作为工业城市高速发展的历史，是一代天津人的记忆。地块周边有文化中心、梅江会展中心、梅江与梅江南居住区、海河后五公里等重要功能区，区域未来将要顺应天津市城市重心南移的发展趋势，为文化中心周边区域外溢的产业功能和居住需求提供容纳对接场所，由工业区转变为居住、商业为主的混合区域。

无界景观工作室与天津市城市规划设计研究院、天津愿景城市开发与设计策划有限公司合作，完成了天津解放南路地区的景观专项规划。随后，无界设计团队又对场地内的中央绿洲（都市绿洲）城市公园及西南端的起步区公园进行了景观设计。

天津解放南路区位图

01 天津解放南路地区景观专项规划
Tianjin Jiefang South Road Landscape Planning

项目地点： 中国 天津市　　**Project location:** Tianjin, China
项目规模： 16.7 公顷　　　**Project scale:** 16.7 hectare
设计时间： 2011—2014 年　 **Design period:** 2011-2014

景观系统专项规划希望针对场地的区位、自然与文化要素，搭建具有内部连接性的城市自然区域和开放空间的网络，构成保证环境、社会与经济可持续发展的生态框架，以多行业、跨学科的多元参与性和实际操作性，促进地块由生产向生活的转变，促进地区的生态改善，促进多元文化的交融，连接城市的过去、现在与未来。

设计团队将绿色基础设施理念融入规划设计概念之中，将绿色公共空间作为区域融合的媒介，在提高景观视觉价值的同时改善区域环境、提升市民幸福感。将公园与周边市政道路一体化设计，调整非机动车道位置，拓宽机非分隔带，将人行道移入公园内，创造舒适的林荫环境；鼓励在公园中穿行、娱乐、休闲，使绿地复合多元功能、融入市民生活，建设大都市中的"乐活公园"。

结合现状条件及解放南路地区生态宜居社区的规划定位，设计团队提出景观规划"生态、文化、乐活"三大主题。

"生态"是指打开公园的边界，打造融入都市的绿色氧吧，配合生态、交通、地下空间等专项规划，建设节约型园林，作为区域融合的媒介。

"文化"是指体现场所特征，突出景观的地域特色，尊重现有自然及历史文脉，保留与活化工业遗存，传承天津独特的市民文化。

"乐活"则是倡导健康、自然、低碳的"乐活"理念，激发活力场景的不断变换，促进居民行为的自然转化，培养健康、可持续的生活方式，延续并创造城市记忆。

设计团队创新性地采用了双重规划体系与二元控制方法，通过四大部分达到对解放南路地区景观系统多层次、全方位的有效控制。景观规划总则部分主要是对现状及相关规划条件的梳理，提出规划原则及策略，构建公共空间景观体系；景观规划通则部分从空间属性及景观元素两个不同角度提出公共空间的景观控制要求，强调景观设计与城市功能的契合，兼顾全区景观风貌的整体性与特色性；概念方案部分对主要公园绿地及道路景观提出控制要求；景观规划细则部分对居住区及公共设施区提出指导细则。❶

---

❶ 参考资料：一体化的绿色基础设施——天津解放南路地区景观规划及公园设计 [J]，住区，2015（3）。

02 天津解放南路地区中央绿轴（都市绿洲）城市公园设计
Tianjin Jiefang South Road Central Green Axis (Urban Oasis) Urban Park Design

- 项目地点：中国 天津市
- 项目规模：100.2 公顷
- 设计时间：2011—2014 年

- **Project location:** Tianjin, China
- **Project scale:** 100.2 hectare
- **Design period:** 2011-2014

都市绿洲是解放南路地区的中央绿轴，南北长3.6公里，东西宽150米。相对于一般城市公园，带状公园与城市的交界面更长，关系更为紧密。

都市绿洲设计以建设"连接生态文明的复合型带状公园、城镇型绿道的样本"为目标，延续解放南路地区景观专项规划提出的"生态、文化、乐活"的定位，提取场地的文化基因，复合休闲健身、文化娱乐等多种功能，增进公园、城市、市民生活三者之间的互动，打造全区的中心绿轴，实现中央绿轴公园的复合功能与多重价值。对工业遗存采取"弃而不废"的原则，最大限度地保留相关遗存，将其改造成都市生活的休闲场所，促进地区的生态改善，延续地域的特色景观。

通过一体化的布局统筹使得带状公园消解于城市，采用一体化的表达方式营造有归属感的公共空间，运用一体化的生态措施高效提升整体环境舒适度。通过多个层面的景观统筹使公园与城市融为一体，连接城市绿道、风道系统、绿色空间等市政、公共服务设施以及工业遗存、市民生活，使公园最终成为引领"绿色新生活"的场所与媒介，鼓励绿色出行、全民健身、创意文化，使资源节约、环境友好等观念深入人心❶。

❶ 参考资料：《中国建筑艺术年鉴》(2013—2014年)。

天津解放南路都市绿洲区位

都市绿洲鸟瞰图

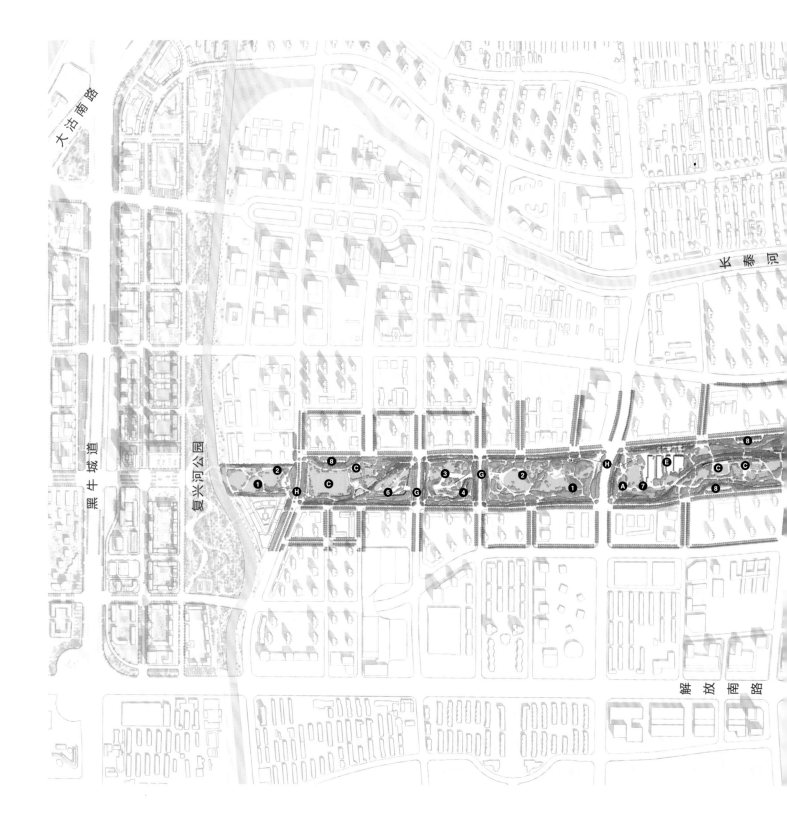

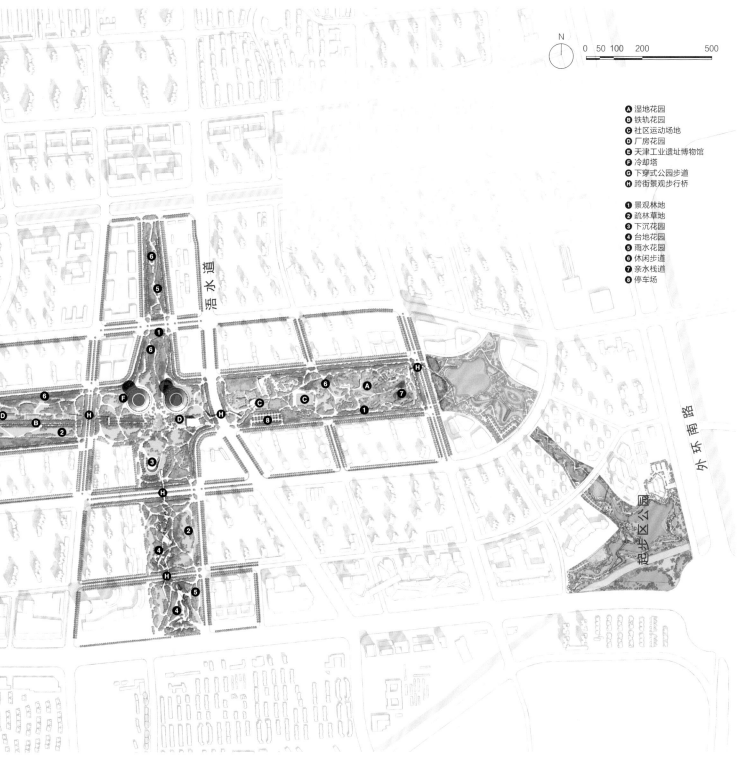

都市绿洲总平面图

（1）节约高效的一体化生态措施——营造舒适的小气候

城市的快速发展使得天津不可避免地遇到了多种生态问题：空气污染、水资源紧缺、土壤盐碱化、生物多样性缺乏以及能源紧缺等，城市生态环境脆弱。原有的工业设施造成了现状区域内的污染、水质及土质均较差等问题。如何使公园为解放南路地区及天津市发挥更大的生态效益，是景观设计团队必须要思考的问题。

紧密结合场地的自然条件，综合微地形、水体、植物等元素，设计团队提出具有针对性、整体性、可持续性的一体化生态措施。通过统筹区域内通风廊道建设、微地形与植被设计，营造舒适的小气候，提升环境舒适度；综合雨水收集、能源利用等生态措施，建立低影响开发雨水系统，让城市水文过程重回自然；建设节约型园林，发挥科普教育、宣传示范作用。

在公园内形成了四种环境较为舒适的区域。① 夏季荫凉通风的区域：利用微地形、景观构筑物和植物，在一定程度上改变气流方向与强度；② 冬季温暖的区域：利用微地形、挡墙等遮挡冬季寒冷西北风，场地中的慢跑径与地热能相结合，营造利于植物生长及室外活动的温暖空间；③ 湿润的区域：喷泉、冷雾等水景与雨水收集系统结合，增湿、降温、除尘，提升环境舒适度；④ 高负氧离子的区域：注重群落品种搭配和单元规模，提升空气中的负氧离子含量和空气质量。

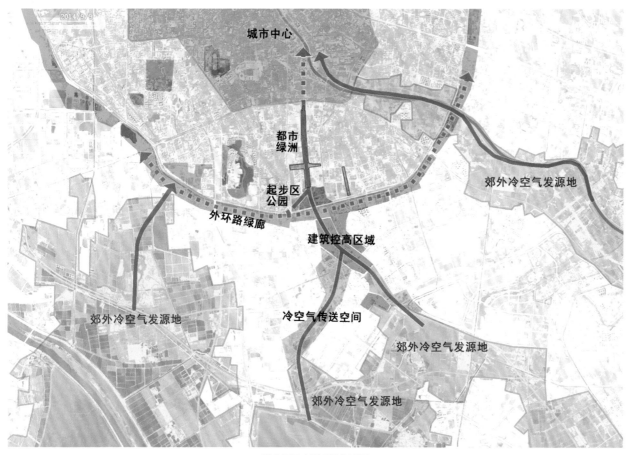

城市通风廊道系统分析图

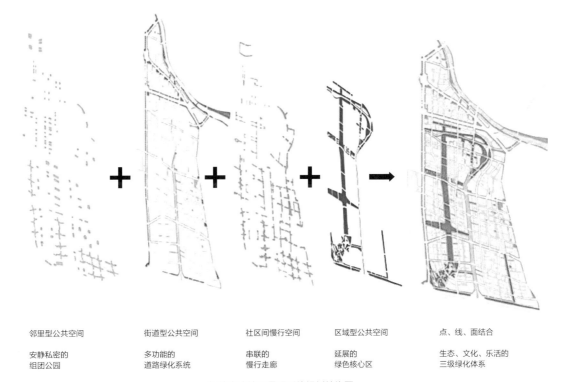

邻里型公共空间	街道型公共空间	社区间慢行空间	区域型公共空间	点、线、面结合
安静私密的组团公园	多功能的道路绿化系统	串联的慢行走廊	延展的绿色核心区	生态、文化、乐活的三级绿化体系

解放南路地区景观系统规划结构图

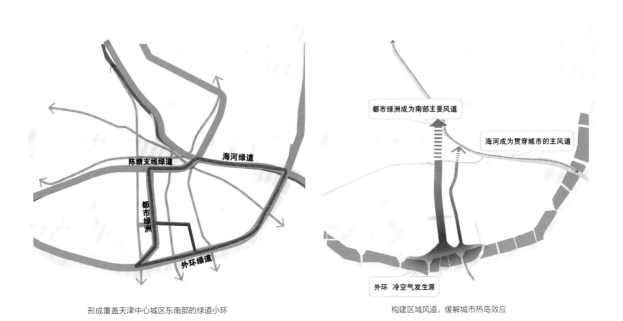

形成覆盖天津中心城区东南部的绿道小环　　构建区域风道，缓解城市热岛效应

都市绿洲与城市的关系

161

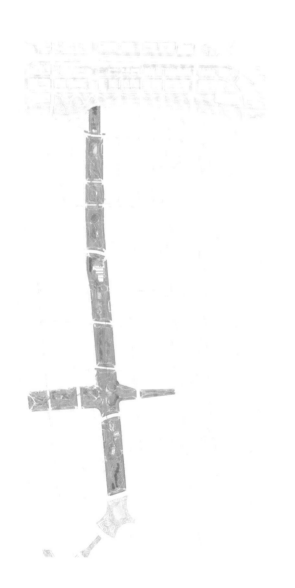

| 夏季荫凉通风的区域 | 冬季温暖的区域 |
| 湿润的区域 | 高负氧离子的区域 |

营造舒适的小气候

· 通风廊道系统设计

天津城市空气不流通现象严重，只有5级以上的大风才能吹进城市。大风时城区飞沙走石、尘土飞扬；弱风时城区热岛效应明显，空气相对静止、异常闷热。空气不流通还使得工矿及交通工具等排放的有害气体在城区沉积，危及居民健康。针对上述问题，项目所在区域的绿洲通风廊道构建的主要目标是加强两方面的空气流通：一是城市与郊区之间的大循环，结合城市弱风主导方向，将郊外的冷空气引入城区，形成局地环流，从而缓解城市热岛效应，并将空气中的有害物质带离城区；二是绿地与周边地块之间的微循环，加强空气交换与自净，减少浮尘。

通风廊道由边缘林地空间、过渡空间、中心开敞楔形空间三个部分构成。边缘林地空间主要通过水平和垂直方向上多层次的植物空间有效吸附浮尘；过渡空间则进一步发挥植被组合的吸附、净化功能；中心开敞楔形空间作为主要的空气通道，结合疏林草地、水体等，使空气在流通过程中也能得到进一步的降温和除尘。

"都市绿洲"带状绿地通风廊道的设计使天津南部片区多了一条会呼吸的通道，并且能像人类的"肺"一样实现对空气的净化功能，改善片区环境，为创造舒适宜人的室外活动空间打下基础。

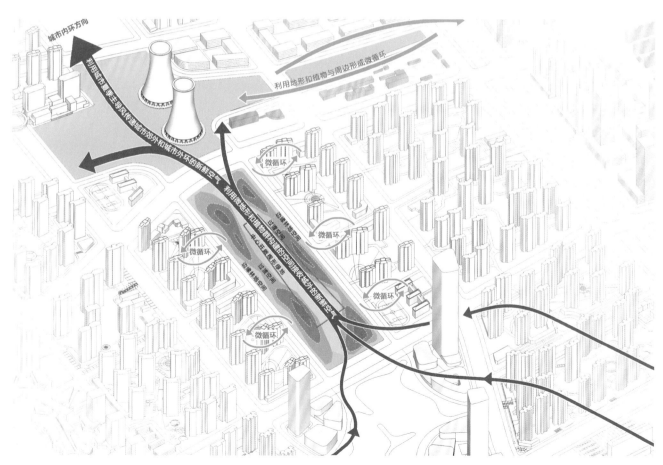

城市通风廊道
空气流动模式图

· **低影响开发雨水系统设计**

北方地区夏季暴雨频发，城市现有灰色基础设施应对能力有限。天津市年平均降水量在360～970毫米，平均值600毫米上下（1949—2010年数据），其中75%的降水集中在6～8月。以公园绿地作为吸纳暴雨径流的巨大"海绵体"，是本项目雨洪利用与景观一体化的尝试。

景观设计遵循上位规划和住建部《海绵城市建设技术指南》要求，通过对雨水径流实施调蓄、净化和利用，改善城市水环境和生态环境；通过各种人工或自然渗透设施使雨水渗入地下，补充地下水资源。本项目中"低影响开发系统"主要处理公园绿地及相邻市政道路的雨水，主要功能为削减城市径流污染负荷、节约水资源。

根据项目地块被割裂的现状和雨水污染源不同的情况，采取两种策略。一是分区域处理，强调就地疏解、局部消纳；节约建设成本，就近与市政雨水管网衔接；化整为零，在各个地块内部完成收集利用。二是分源头收集，对城市道路与公园绿地的径流分别收集处理；利用收集净化的雨水打造公园内的水景，体现节水理念的同时，传承天津因水而生的水文化，激发城市活力。

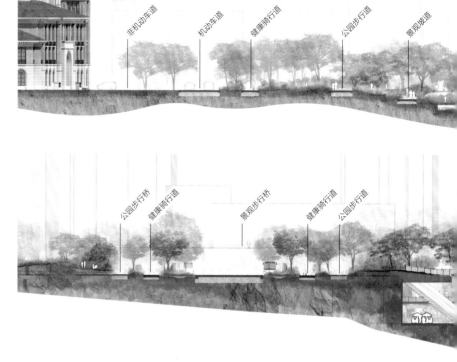

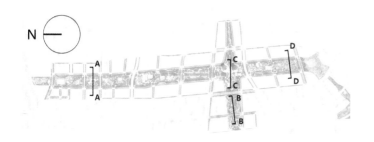

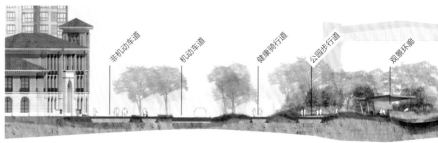

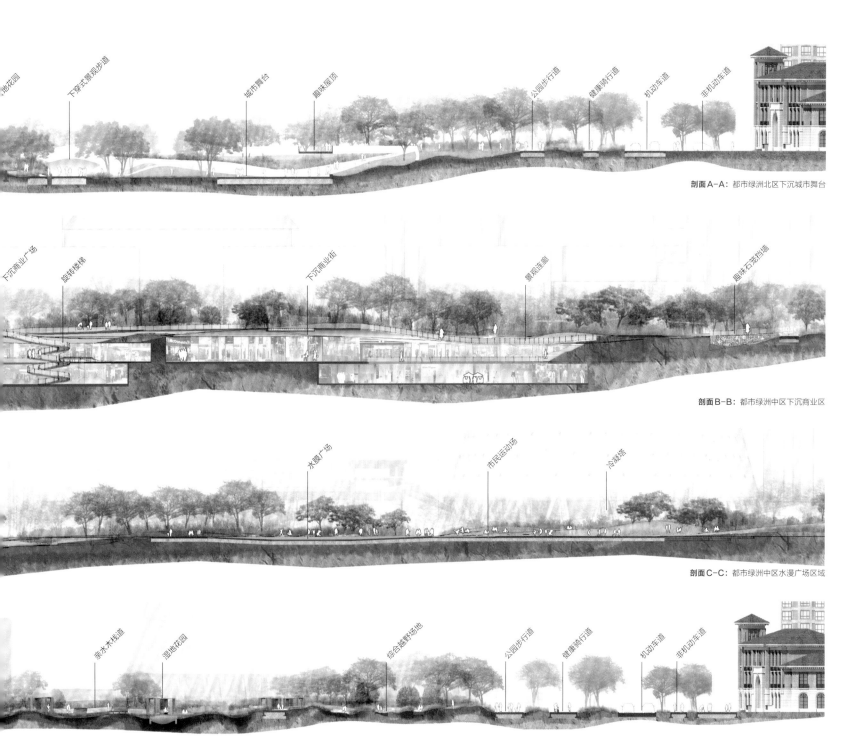

剖面A-A： 都市绿洲北区下沉城市舞台

剖面B-B： 都市绿洲中区下沉商业区

剖面C-C： 都市绿洲中区水漫广场区域

剖面D-D： 都市绿洲南区一期实施地块湿地花园

都市绿洲剖面图

（2）突出场所特征的一体化文化表达

天津是依托海河发展起来的移民城市和商埠城市。有"卫嘴子"之称的天津人乐观、豁达、知足，具有独特的"乐活"精神。同时，近代的工业发展在天津留下大量印记，场地内的热电厂工业遗存是一代人的记忆。设计团队利用场地内的工业遗存，以设计展现天津人的乐天气质、娱乐精神，增进居民对于环境的认同感和归属感。

此外，天津作为北方城市，拥有独特的水文化，同时又面临水资源匮乏的现状。在设计中用技术手段对水资源进行节约和利用，体现了天津节水型城市的特点。

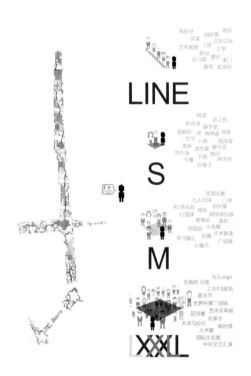

不同尺度的丰富空间

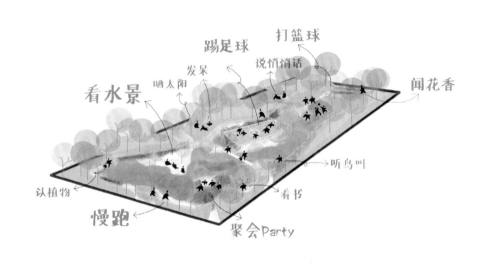

体现天津人乐观、豁达、知足，具有独特的"乐活"精神（Enjoy Dailylife）

将热电厂冷却塔改造为活力地标,结合现状鱼塘打造湿地花园,利用人造雾技术形成梦幻的彩虹景观

（3）公园与周边道路一体化设计

以步行为主的慢行交通是城市空间框架的最佳载体，它在承载日常生活、展示城市形象、促进文化交流、提升空间品质、聚集中心人气、组织内外交通方面，具有不可替代的作用。

将公园与市政道路一体化设计，通过调整规划的市政道路断面，加宽机动车与非机动车道之间的绿化隔离带，使带状公园融入城市绿道系统，每一个建设环节都对城市慢行交通有所贡献，让绿色出行体验更为安全、舒适、愉悦。

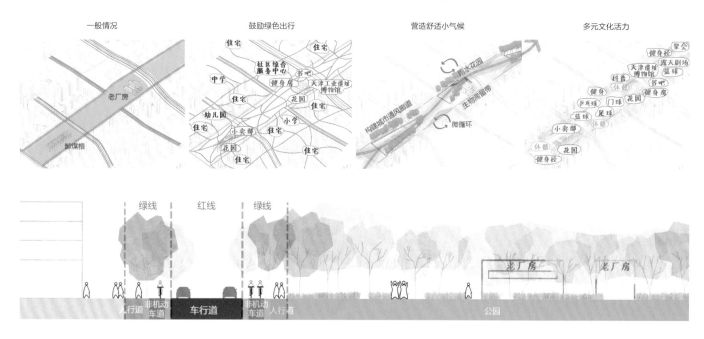

公园与周边道路一体化策略

03 天津解放南路地区起步区公园景观设计
Tianjin Jiefang South Road Pilot Area Park Design

项目地点： 中国 天津市
项目规模： 17.7公顷
设计时间： 2011—2014年

Project location: Tianjin, China
Project scale: 17.7hectare
Design period: 2011-2014

起步区公园位于天津解放南路地区西南端，以解放南路和外环线的交叉点为起点，向卫津河、太湖路方向展开，占地面积约17.7公顷，于2014年7月建成并投入使用。起步区公园所在区域是天津市区的南大门，是城市入口对外形象展示的窗口区域，园内现代工业标志性景观构筑物能激发工业记忆的联想。

结合解放南路地区生态宜居社区的定位，景观设计团队充分利用区域现状丰富的水资源，以及园内具有现代工业美感的标志性景观构筑物与绿色自然景观基底形成的戏剧化反差，营造亲近自然的、延续地域文化特征的环境，创造赏心悦目、复合多种功能的城市公园。

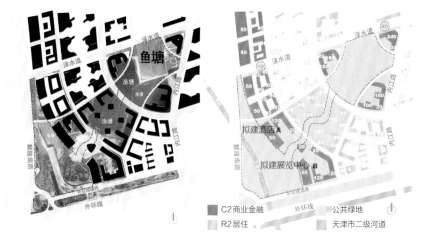

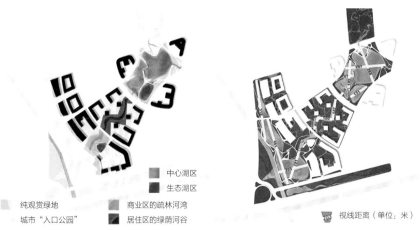

天津解放南路地区起步区公园分析图

（1）注重景观与建筑的统筹安排，塑造城市入口新形象

起步区公园的地块现状为大面积的鱼塘水系与湖堤地貌，有少量的简易构筑物，周边现状为老旧的工业、市场、居民小区等。根据城市总体规划和解放南路地区相关规划要求，地块周边未来布局为商业办公建筑和居民住宅，并设置地铁站点。

起步区公园由位于太湖路两侧及卫津河上的三个公园组成，构成"一河连两湖"的景观结构，为周边的商业金融建筑和中高端住宅社区所环绕。公园的景观设计从"亲水亲绿"的理念出发，充分利用现状丰富的水资源和湖堤形态，作为水型设计的基本形态与出发点，减少景观施工土方工程量，通过蜿蜒的自然河道景观、层次丰富的湖景、视野开阔的大湖景观，营造亲近自然的宜人场所。在生态自然的环境上叠加艺术化、人性化的景观元素，使得水系、驳岸、景观、建筑交相呼应，创造优美、轻松、绿色的景观门户。公园与周边公共建筑共同构成天津城新的形象入口，面向城市展开一幅优美的画卷。

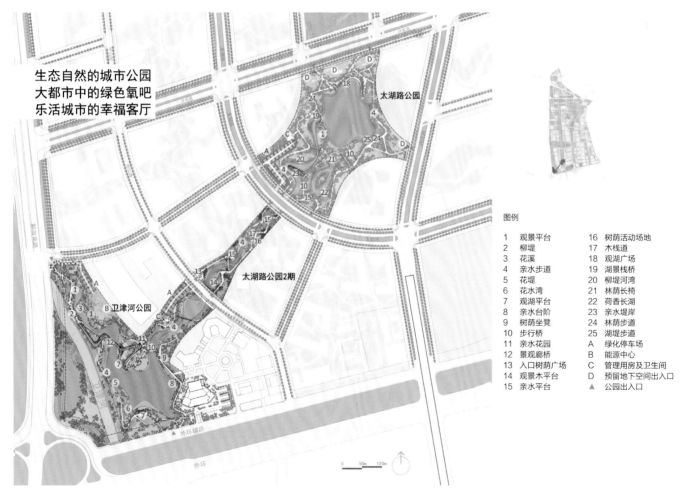

天津解放南路地区起步区公园平面图

施工中的起步区公园

起步区公园湖景

（2）注重复合多种功能，构建一体化的绿色基础设施

配合生态、交通、地下空间等专项规划研究，设计团队提出了构建多层次、多功能一体化绿色基础设施的规划设计方案，统筹安排地上地下的园林绿化与市政基础设施，注重资源的合理高效利用。

卫津河公园中部结合微地形设计蜿蜒起伏的景观廊桥，临水设计高挑楼台，现代的廊桥设计与自然野趣的公园景观形成鲜明对比，成为公园中心的视觉焦点。结合地铁站点规划预留地下空间出入口，未来可布局多功能的竖向空间。建立低影响开发雨水系统，形成吸纳暴雨径流的"城市海绵"，对雨水径流实施调蓄、净化和利用，补充地下水资源，改善区域水环境。

园内多功能平台

（3）注重竖向空间的统筹安排，凸显公园的生态节能特色

天津市地热资源十分丰富。解放南路区域处于王兰庄地热田，浅层和深层地热资源条件都较好，属于地埋管地源热泵较适宜区。通过在项目公园的绿地及水系下埋设布置浅层地源热泵系统（埋设地埋管换热器），可解决周边商业办公建筑中央空调系统冬季供热、夏季供冷的需要，不仅为公共建筑提供冬暖夏凉的舒适环境，而且节约大量的能源。公园的景观设计从竖向空间统筹安排的角度出发，充分利用现状鱼塘低洼地貌设置水系，保留现状植物。这种结合现状地貌造景的做法为地源热泵施工提供了良好的条件，同时满足公园景观和生态节能两方面的设计需要。

建成的天津解放南路起步区公园作为综合、立体的城市绿色基础设施，拥有生态化的自然环境和艺术化的景观构筑。这座"乐活"公园改善了城市环境，提升了市民的幸福感，并潜移默化地影响着人们的生活方式，促进人与自然的良性发展。

公园内的步行桥

起步区公园秋景

后　　记

　　无界景观工作室成立于2004年，一直在城市更新、乡村振兴、城乡融合、生态修复、文旅发展、历史文化保护与传承等领域，致力于探索景观设计融合多学科智慧、统筹多专业协作的最行之有效的途径与方法，助力社会、经济和生态的绿色可持续发展。

　　工作室始终关注人的日常生活与公共空间的联系，关注景观设计对社会关系的良性引导，对生态系统的保护修复，探寻因地、因时、因人而异的，与同质化相悖的解决方案；始终坚持以专业手段协调人与环境的关系，将风景融入日常生活，缓解人的生存压力，激发民众活力与场地生产力，提升"安住"者的幸福感与归属感。

　　格物而致知，笃行而致远。未来，愿行而不辍，继续为城、为乡、为人、为生境，营建整体的、连续的美丽。

AFTERWORD

Founded in 2004, View Unlimited Landscape Architecture Studio has been dedicated to exploring the most effective ways and methods for integrating multidisciplinary ideas and coordinating multi-disciplinary collaboration in the fields of urban renewal, rural revitalization, urban-rural integration, ecological restoration, cultural tourism and development, and historical and cultural preservation and inheritance, to support the green and sustainable development of our society, economy and ecology.

We focus on the inherent connection between people's daily life and public spaces, with an emphasis on the constructive integration of landscape design into social relationships, the protection and restoration of the ecosystem, and the search for solutions that veer away from homogenization by taking into account the surrounding environment, time period, and community. We strive to harmonize the relationship between people and the environment by employing professional techniques and integrating landscapes into everyday life that relieve the pressures of human existence, stimulate people's vitality and productivity, and ultimately enhance a sense of happiness and belonging in those who simply wish to "abide".

Attaining knowledge from our origins, our actions echo far beyond. Our aim is to persevere in creating a holistic and sustainable aesthetic for cities, towns, people, and our natural habitat.

致谢

感谢著名建筑师朱小地先生和崔愷院士的推荐,使我们有幸参与北京城市副中心先行启动区景观设计项目,亦感谢两位先生在项目实践过程中给予的专业指导。

感谢王得后老师、赵园老师在北京文化解读方面不吝赐教,为项目设计寻脉溯源。

感谢黄海涛老师在文化艺术方面给予的专业指导。

感谢瞿志老师在工程技术方面给予的专业指导。

感谢所有对本项目给予支持与帮助的部门领导、专家学者和热心市民。

感谢所有参与本项目的设计师和工作人员。